Linear Matrix Inequalities in System and Control Theory

SIAM Studies in Applied and Numerical Mathematics

This series of monographs focuses on mathematics and its applications to problems of current concern to industry, government, and society. These monographs will be of interest to applied mathematicians, numerical analysts, statisticians, engineers, and scientists who have an active need to learn useful methodology.

Series List

Vol. 1 *Lie-Bäcklund Transformations in Applications*
Robert L. Anderson and Nail H. Ibragimov

Vol. 2 *Methods and Applications of Interval Analysis*
Ramon E. Moore

Vol. 3 *Ill-Posed Problems for Integrodifferential Equations in Mechanics and Electromagnetic Theory*
Frederick Bloom

Vol. 4 *Solitons and the Inverse Scattering Transform*
Mark J. Ablowitz and Harvey Segur

Vol. 5 *Fourier Analysis of Numerical Approximations of Hyperbolic Equations*
Robert Vichnevetsky and John B. Bowles

Vol. 6 *Numerical Solution of Elliptic Problems*
Garrett Birkhoff and Robert E. Lynch

Vol. 7 *Analytical and Numerical Methods for Volterra Equations*
Peter Linz

Vol. 8 *Contact Problems in Elasticity: A Study of Variational Inequalities and Finite Element Methods*
N. Kikuchi and J. T. Oden

Vol. 9 *Augmented Lagrangian and Operator-Splitting Methods in Nonlinear Mechanics*
Roland Glowinski and P. Le Tallec

Vol. 10 *Boundary Stabilization of Thin Plate Splines*
John E. Lagnese

Vol. 11 *Electro-Diffusion of Ions*
Isaak Rubinstein

Vol. 12 *Mathematical Problems in Linear Viscoelasticity*
Mauro Fabrizio and Angelo Morro

Vol. 13 *Interior-Point Polynomial Algorithms in Convex Programming*
Yurii Nesterov and Arkadii Nemirovskii

Vol. 14 *The Boundary Function Method for Singular Perturbation Problems*
Adelaida B. Vasil'eva, Valentin F. Butuzov, and Leonid V. Kalachev

Vol. 15 *Linear Matrix Inequalities in System and Control Theory*
Stephen Boyd, Laurent El Ghaoui, Eric Feron, and Venkataramanan Balakrishnan

Vol. 16 *Indefinite-Quadratic Estimation and Control: A Unified Approach to H^2 and H^∞ Theories*
Babak Hassibi, Ali H. Sayed, and Thomas Kailath

Stephen Boyd
Stanford University
Stanford, California

Eric Feron
Massachusetts Institute of Technology
Cambridge, Massachusetts

Laurent El Ghaoui
University of California,
Berkeley, California

Venkataramanan Balakrishnan
Purdue University
West Lafayette, Indiana

Linear Matrix Inequalities in System and Control Theory

siam

Society for Industrial and Applied Mathematics
Philadelphia

The royalties from the sales of this book are being placed in a fund to help students attend SIAM meetings and other SIAM related activities. This fund is administered by SIAM and qualified individuals are encouraged to write directly to SIAM for guidelines.

10 9 8 7 6 5 4

Library of Congress Cataloging-in-Publication Data

Linear matrix inequalities in system and control theory / Stephen Boyd
... [et al.].
p. cm. -- (SIAM studies in applied mathematics ; vol. 15)
Includes bibliographical references and index.
ISBN 0-89871-485-0
1. Control theory. 2. Matrix inequalities. 3. Mathematical optimization. I. Boyd, Stephen P. II. Series: SIAM studies in applied mathematics : 15.
QA402.3.L489 1994
515'.64--dc20 94-10477

Contents

Preface

The basic topic of this book is solving problems from system and control theory using convex optimization. We show that a wide variety of problems arising in system and control theory can be reduced to a handful of standard convex and quasiconvex optimization problems that involve matrix inequalities. For a few special cases there are "analytic solutions" to these problems, but our main point is that they can be solved numerically in all cases. These standard problems can be solved in polynomial-time (by, e.g., the ellipsoid algorithm of Shor, Nemirovskii, and Yudin), and so are tractable, at least in a theoretical sense. Recently developed interior-point methods for these standard problems have been found to be extremely efficient in practice. Therefore, we consider the original problems from system and control theory as solved.

This book is primarily intended for the researcher in system and control theory, but can also serve as a source of application problems for researchers in convex optimization. Although we believe that the methods described in this book have great practical value, we should warn the reader whose primary interest is applied control engineering. This is a research monograph: We present no specific examples or numerical results, and we make only brief comments about the implications of the results for practical control engineering. To put it in a more positive light, we hope that this book will later be considered as the *first* book on the topic, not the most readable or accessible.

The background required of the reader is knowledge of basic system and control theory and an exposure to optimization. Sontag's book *Mathematical Control Theory* [SON90] is an excellent survey. Further background material is covered in the texts *Linear Systems* [KAI80] by Kailath, *Nonlinear Systems Analysis* [VID92] by Vidyasagar, *Optimal Control: Linear Quadratic Methods* [AM90] by Anderson and Moore, and *Convex Analysis and Minimization Algorithms I* [HUL93] by Hiriart–Urruty and Lemaréchal.

We also highly recommend the book *Interior-point Polynomial Algorithms in Convex Programming* [NN94] by Nesterov and Nemirovskii as a companion to this book. The reader will soon see that their ideas and methods play a critical role in the basic idea presented in this book.

Acknowledgments

We thank A. Nemirovskii for encouraging us to write this book. We are grateful to S. Hall, M. Gevers, and A. Packard for introducing us to multiplier methods, realization theory, and state-feedback synthesis methods, respectively. This book was greatly improved by the suggestions of J. Abedor, P. Dankoski, J. Doyle, G. Franklin, T. Kailath, R. Kosut, I. Petersen, E. Pyatnitskii, M. Rotea, M. Safonov, S. Savastiouk, A. Tits, L. Vandenberghe, and J. C. Willems. It is a special pleasure to thank V. Yakubovich for his comments and suggestions.

The research reported in this book was supported in part by AFOSR (under F49620-92-J-0013), NSF (under ECS-9222391), and ARPA (under F49620-93-1-0085). L. El Ghaoui and E. Feron were supported in part by the Délégation Générale pour l'Armement. E. Feron was supported in part by the Charles Stark Draper Career Development Chair at MIT. V. Balakrishnan was supported in part by the Control and Dynamical Systems Group at Caltech and by NSF (under NSFD CDR-8803012).

This book was typeset by the authors using LaTeX, and many macros originally written by Craig Barratt.

Stephen Boyd — *Stanford, California*
Laurent El Ghaoui — *Paris, France*
Eric Feron — *Cambridge, Massachusetts*
Venkataramanan Balakrishnan — *College Park, Maryland*

Linear Matrix Inequalities in System and Control Theory

Chapter 1

Introduction

1.1 Overview

The aim of this book is to show that we can reduce a very wide variety of problems arising in system and control theory to a few standard convex or quasiconvex optimization problems involving linear matrix inequalities (LMIs). Since these resulting optimization problems can be solved *numerically* very efficiently using recently developed interior-point methods, our reduction constitutes a solution to the original problem, certainly in a practical sense, but also in several other senses as well. In comparison, the more conventional approach is to seek an analytic or frequency-domain solution to the matrix inequalities.

The types of problems we consider include:

- matrix scaling problems, e.g., minimizing condition number by diagonal scaling
- construction of quadratic Lyapunov functions for stability and performance analysis of linear differential inclusions
- joint synthesis of state-feedback and quadratic Lyapunov functions for linear differential inclusions
- synthesis of state-feedback and quadratic Lyapunov functions for stochastic and delay systems
- synthesis of Lur'e-type Lyapunov functions for nonlinear systems
- optimization over an affine family of transfer matrices, including synthesis of multipliers for analysis of linear systems with unknown parameters
- positive orthant stability and state-feedback synthesis
- optimal system realization
- interpolation problems, including scaling
- multicriterion LQG/LQR
- inverse problem of optimal control

In some cases, we are describing known, published results; in others, we are extending known results. In many cases, however, it seems that the results are new.

By scanning the list above or the table of contents, the reader will see that Lyapunov's methods will be our main focus. Here we have a secondary goal, beyond showing that many problems from Lyapunov theory can be cast as convex or quasiconvex problems. This is to show that Lyapunov's methods, which are traditionally applied to the analysis of system *stability*, can just as well be used to find bounds on system *performance,* provided we do not insist on an "analytic solution".

1.2 A Brief History of LMIs in Control Theory

The history of LMIs in the analysis of dynamical systems goes back more than 100 years. The story begins in about 1890, when Lyapunov published his seminal work introducing what we now call Lyapunov theory. He showed that the differential equation

$$\frac{d}{dt}x(t) = Ax(t) \tag{1.1}$$

is stable (i.e., all trajectories converge to zero) if and only if there exists a positive-definite matrix P such that

$$A^TP + PA < 0. \tag{1.2}$$

The requirement $P > 0$, $A^TP + PA < 0$ is what we now call a Lyapunov inequality on P, which is a special form of an LMI. Lyapunov also showed that this first LMI could be explicitly solved. Indeed, we can pick any $Q = Q^T > 0$ and then solve the linear equation $A^TP + PA = -Q$ for the matrix P, which is guaranteed to be positive-definite if the system (1.1) is stable. In summary, the first LMI used to analyze stability of a dynamical system was the Lyapunov inequality (1.2), which can be solved analytically (by solving a set of linear equations).

The next major milestone occurs in the 1940's. Lur'e, Postnikov, and others in the Soviet Union applied Lyapunov's methods to some specific practical problems in control engineering, especially, the problem of stability of a control system with a nonlinearity in the actuator. Although they did not explicitly form matrix inequalities, their stability criteria have the form of LMIs. These inequalities were reduced to polynomial inequalities which were then checked "by hand" (for, needless to say, small systems). Nevertheless they were justifiably excited by the idea that Lyapunov's theory could be applied to important (and difficult) practical problems in control engineering. From the introduction of Lur'e's 1951 book [LUR57] we find:

> This book represents the first attempt to demonstrate that the ideas expressed 60 years ago by Lyapunov, which even comparatively recently appeared to be remote from practical application, are now about to become a real medium for the examination of the urgent problems of contemporary engineering.

In summary, Lur'e and others were the first to apply Lyapunov's methods to practical control engineering problems. The LMIs that resulted were solved analytically, by hand. Of course this limited their application to small (second, third order) systems.

The next major breakthrough came in the early 1960's, when Yakubovich, Popov, Kalman, and other researchers succeeded in reducing the solution of the LMIs that arose in the problem of Lur'e to simple graphical criteria, using what we now call the positive-real (PR) lemma (see §2.7.2). This resulted in the celebrated Popov criterion, circle criterion, Tsypkin criterion, and many variations. These criteria could be applied to higher order systems, but did not gracefully or usefully extend to systems containing more than one nonlinearity. From the point of view of our story (LMIs in control theory), the contribution was to show how to solve a certain family of LMIs by graphical methods.

The important role of LMIs in control theory was already recognized in the early 1960's, especially by Yakubovich [YAK62, YAK64, YAK67]. This is clear simply from the titles of some of his papers from 1962–5, e.g., *The solution of certain matrix inequalities in automatic control theory* (1962), and *The method of matrix inequalities in the stability theory of nonlinear control systems* (1965; English translation 1967).

The PR lemma and extensions were intensively studied in the latter half of the 1960s, and were found to be related to the ideas of passivity, the small-gain criteria introduced by Zames and Sandberg, and quadratic optimal control. By 1970, it was known that the LMI appearing in the PR lemma could be solved not only by graphical means, but also by solving a certain algebraic Riccati equation (ARE). In a 1971 paper [WIL71B] on quadratic optimal control, J. C. Willems is led to the LMI

$$\begin{bmatrix} A^TP+PA+Q & PB+C^T \\ B^TP+C & R \end{bmatrix} \geq 0, \tag{1.3}$$

and points out that it can be solved by studying the symmetric solutions of the ARE

$$A^TP+PA-(PB+C^T)R^{-1}(B^TP+C)+Q=0,$$

which in turn can be found by an eigendecomposition of a related Hamiltonian matrix. (See §2.7.2 for details.) This connection had been observed earlier in the Soviet Union, where the ARE was called the Lur'e resolving equation (see [YAK88]).

So by 1971, researchers knew several methods for solving special types of LMIs: direct (for small systems), graphical methods, and by solving Lyapunov or Riccati equations. From our point of view, these methods are all "closed-form" or "analytic" solutions that can be used to solve special forms of LMIs. (Most control researchers and engineers consider the Riccati equation to have an "analytic" solution, since the standard algorithms that solve it are very predictable in terms of the effort required, which depends almost entirely on the problem size and not the particular problem data. Of course it cannot be solved exactly in a finite number of arithmetic steps for systems of fifth and higher order.)

In Willems' 1971 paper we find the following striking quote:

> The basic importance of the LMI seems to be largely unappreciated. It would be interesting to see whether or not it can be exploited in computational algorithms, for example.

Here Willems refers to the specific LMI (1.3), and not the more general form that we consider in this book. Still, Willems' suggestion that LMIs might have some advantages in computational algorithms (as compared to the corresponding Riccati equations) foreshadows the next chapter in the story.

The next major advance (in our view) was the simple observation that:

> The LMIs that arise in system and control theory can be formulated as *convex optimization problems* that are amenable to computer solution.

Although this is a simple observation, it has some important consequences, the most important of which is that we can reliably solve many LMIs for which no "analytic solution" has been found (or is likely to be found).

This observation was made explicitly by several researchers. Pyatnitskii and Skorodinskii [PS82] were perhaps the first researchers to make this point, clearly and completely. They reduced the original problem of Lur'e (extended to the case of multiple nonlinearities) to a convex optimization problem involving LMIs, which they then solved using the ellipsoid algorithm. (This problem had been studied before, but the "solutions" involved an arbitrary scaling matrix.) Pyatnitskii and Skorodinskii were the first, as far as we know, to formulate the search for a Lyapunov function as a convex optimization problem, and then apply an algorithm guaranteed to solve the optimization problem.

We should also mention several precursors. In a 1976 paper, Horisberger and Belanger [HB76] had remarked that the existence of a quadratic Lyapunov function that simultaneously proves stability of a collection of linear systems is a convex problem involving LMIs. And of course, the idea of having a computer search for a Lyapunov function was not new—it appears, for example, in a 1965 paper by Schultz et al. [SSHJ65].

The final chapter in our story is quite recent and of great practical importance: the development of powerful and efficient interior-point methods to solve the LMIs that arise in system and control theory. In 1984, N. Karmarkar introduced a new linear programming algorithm that solves linear programs in polynomial-time, like the ellipsoid method, but in contrast to the ellipsoid method, is also very efficient in practice. Karmarkar's work spurred an enormous amount of work in the area of interior-point methods for linear programming (including the rediscovery of efficient methods that were developed in the 1960s but ignored). Essentially all of this research activity concentrated on algorithms for linear and (convex) quadratic programs. Then in 1988, Nesterov and Nemirovskii developed interior-point methods that apply directly to convex problems involving LMIs, and in particular, to the problems we encounter in this book. Although there remains much to be done in this area, several interior-point algorithms for LMI problems have been implemented and tested on specific families of LMIs that arise in control theory, and found to be extremely efficient.

A summary of key events in the history of LMIs in control theory is then:

- **1890:** First LMI appears; analytic solution of the Lyapunov LMI via Lyapunov equation.
- **1940's:** Application of Lyapunov's methods to real control engineering problems. Small LMIs solved "by hand".
- **Early 1960's:** PR lemma gives graphical techniques for solving another family of LMIs.
- **Late 1960's:** Observation that the same family of LMIs can be solved by solving an ARE.
- **Early 1980's:** Recognition that many LMIs can be solved by computer via convex programming.
- **Late 1980's:** Development of interior-point algorithms for LMIs.

It is fair to say that Yakubovich is the father of the field, and Lyapunov the grandfather of the field. The new development is the ability to directly *solve* (general) LMIs.

1.3 Notes on the Style of the Book

We use a very informal mathematical style, e.g., we often fail to mention regularity or other technical conditions. Every statement is to be interpreted as being true modulo appropriate technical conditions (that in most cases are trivial to figure out).

We are very informal, perhaps even cavalier, in our reduction of a problem to an optimization problem. We sometimes skip "details" that would be important if the optimization problem were to be solved numerically. As an example, it may be necessary to add constraints to the optimization problem for normalization or to ensure boundedness. We do not discuss initial guesses for the optimization problems, even though good ones may be available. Therefore, the reader who wishes to *implement* an algorithm that solves a problem considered in this book should be prepared to make a few modifications or additions to our description of the "solution".

In a similar way, we do not pursue any theoretical aspects of reducing a problem to a convex problem involving matrix inequalities. For example, for each reduced problem we could state, probably simplify, and then interpret in system or control theoretic terms the optimality conditions for the resulting convex problem. Another fascinating topic that could be explored is the relation between system and control theory duality and convex programming duality. Once we reduce a problem arising in control theory to a convex program, we can consider various dual optimization problems, lower bounds for the problem, and so on. Presumably these dual problems and lower bounds can be given interesting system-theoretic interpretations.

We mostly consider continuous-time systems, and assume that the reader can translate the results from the continuous-time case to the discrete-time case. We switch to discrete-time systems when we consider system realization problems (which almost always arise in this form) and also when we consider stochastic systems (to avoid the technical details of stochastic differential equations).

The list of problems that we consider is meant only to be representative, and certainly not exhaustive. To avoid excessive repetition, our treatment of problems becomes more terse as the book progresses. In the first chapter on analysis of linear differential inclusions, we describe many variations on problems (e.g., computing bounds on margins and decay rates); in later chapters, we describe fewer and fewer variations, assuming that the reader could work out the extensions.

Each chapter concludes with a section entitled Notes and References, in which we hide proofs, precise statements, elaborations, and bibliography and historical notes. The completeness of the bibliography should not be overestimated, despite its size (over 500 entries). The appendix contains a list of notation and a list of acronyms used in the book. We apologize to the reader for the seven new acronyms we introduce.

To lighten the notation, we use the standard convention of dropping the time argument from the variables in differential equations. Thus, $\dot{x} = Ax$ is our short form for $dx/dt = Ax(t)$. Here A is a constant matrix; when we encounter time-varying coefficients, we will explicitly show the time dependence, as in $\dot{x} = A(t)x$. Similarly, we drop the dummy variable from definite integrals, writing for example, $\int_0^T u^T y\,dt$ for $\int_0^T u(t)^T y(t)\,dt$. To reduce the number of parentheses required, we adopt the convention that the operators $\mathbf{Tr}$ (trace of a matrix) and $\mathbf{E}$ (expected value) have lower precedence than multiplication, transpose, etc. Thus, $\mathbf{Tr}\, A^T B$ means $\mathbf{Tr}\left(A^T B\right)$.

1.4 Origin of the Book

This book started out as a section of the paper *Method of Centers for Minimizing Generalized Eigenvalues,* by Boyd and El Ghaoui [BE93], but grew too large to be a section. For a few months it was a manuscript (that presumably would have been submitted for publication as a paper) entitled *Generalized Eigenvalue Problems Arising in Control Theory.* Then Feron, and later Balakrishnan, started adding material, and soon it was clear that we were writing a book, not a paper. The order of the authors' names reflects this history.

Chapter 2

Some Standard Problems Involving LMIs

2.1 Linear Matrix Inequalities

A linear matrix inequality (LMI) has the form

$$F(x) \stackrel{\Delta}{=} F_0 + \sum_{i=1}^{m} x_i F_i > 0, \tag{2.1}$$

where $x \in \mathbf{R}^m$ is the variable and the symmetric matrices $F_i = F_i^T \in \mathbf{R}^{n \times n}$, $i = 0, \ldots, m$, are given. The inequality symbol in (2.1) means that $F(x)$ is positive-definite, i.e., $u^T F(x) u > 0$ for all nonzero $u \in \mathbf{R}^n$. Of course, the LMI (2.1) is equivalent to a set of n polynomial inequalities in x, i.e., the leading principal minors of $F(x)$ must be positive.

We will also encounter *nonstrict* LMIs, which have the form

$$F(x) \geq 0. \tag{2.2}$$

The strict LMI (2.1) and the nonstrict LMI (2.2) are closely related, but a precise statement of the relation is a bit involved, so we defer it to §2.5. In the next few sections we consider strict LMIs.

The LMI (2.1) is a convex constraint on x, i.e., the set $\{x \mid F(x) > 0\}$ is convex. Although the LMI (2.1) may seem to have a specialized form, it can represent a wide variety of convex constraints on x. In particular, linear inequalities, (convex) quadratic inequalities, matrix norm inequalities, and constraints that arise in control theory, such as Lyapunov and convex quadratic matrix inequalities, can all be cast in the form of an LMI.

Multiple LMIs $F^{(1)}(x) > 0, \ldots, F^{(p)}(x) > 0$ can be expressed as the single LMI $\mathbf{diag}(F^{(1)}(x), \ldots, F^{(p)}(x)) > 0$. Therefore we will make no distinction between a set of LMIs and a single LMI, i.e., "the LMI $F^{(1)}(x) > 0, \ldots, F^{(p)}(x) > 0$" will mean "the LMI $\mathbf{diag}(F^{(1)}(x), \ldots, F^{(p)}(x)) > 0$".

When the matrices F_i are diagonal, the LMI $F(x) > 0$ is just a set of linear inequalities. Nonlinear (convex) inequalities are converted to LMI form using Schur complements. The basic idea is as follows: the LMI

$$\begin{bmatrix} Q(x) & S(x) \\ S(x)^T & R(x) \end{bmatrix} > 0, \tag{2.3}$$

where $Q(x) = Q(x)^T$, $R(x) = R(x)^T$, and $S(x)$ depend affinely on x, is equivalent to

$$R(x) > 0, \qquad Q(x) - S(x)R(x)^{-1}S(x)^T > 0. \tag{2.4}$$

In other words, the set of nonlinear inequalities (2.4) can be represented as the LMI (2.3).

As an example, the (maximum singular value) matrix norm constraint $\|Z(x)\| < 1$, where $Z(x) \in \mathbf{R}^{p\times q}$ and depends affinely on x, is represented as the LMI

$$\begin{bmatrix} I & Z(x) \\ Z(x)^T & I \end{bmatrix} > 0$$

(since $\|Z\| < 1$ is equivalent to $I - ZZ^T > 0$). Note that the case $q = 1$ reduces to a general convex quadratic inequality on x.

The constraint $c(x)^T P(x)^{-1} c(x) < 1$, $P(x) > 0$, where $c(x) \in \mathbf{R}^n$ and $P(x) = P(x)^T \in \mathbf{R}^{n\times n}$ depend affinely on x, is expressed as the LMI

$$\begin{bmatrix} P(x) & c(x) \\ c(x)^T & 1 \end{bmatrix} > 0.$$

More generally, the constraint

$$\mathbf{Tr}\, S(x)^T P(x)^{-1} S(x) < 1, \qquad P(x) > 0,$$

where $P(x) = P(x)^T \in \mathbf{R}^{n\times n}$ and $S(x) \in \mathbf{R}^{n\times p}$ depend affinely on x, is handled by introducing a new (slack) matrix variable $X = X^T \in \mathbf{R}^{p\times p}$, and the LMI (in x and X):

$$\mathbf{Tr}\, X < 1, \qquad \begin{bmatrix} X & S(x)^T \\ S(x) & P(x) \end{bmatrix} > 0.$$

Many other convex constraints on x can be expressed in the form of an LMI; see the Notes and References.

2.1.1 Matrices as variables

We will often encounter problems in which the variables are matrices, e.g., the Lyapunov inequality

$$A^T P + PA < 0, \tag{2.5}$$

where $A \in \mathbf{R}^{n\times n}$ is given and $P = P^T$ is the variable. In this case we will not write out the LMI explicitly in the form $F(x) > 0$, but instead make clear which matrices are the variables. The phrase "the LMI $A^T P + PA < 0$ in P" means that the matrix P is a variable. (Of course, the Lyapunov inequality (2.5) is readily put in the form (2.1), as follows. Let $P_1, \ldots, P_m$ be a basis for symmetric $n \times n$ matrices ($m = n(n+1)/2$). Then take $F_0 = 0$ and $F_i = -A^T P_i - P_i A$.) Leaving LMIs in a condensed form such as (2.5), in addition to saving notation, may lead to more efficient computation; see §2.4.4.

As another related example, consider the quadratic matrix inequality

$$A^T P + PA + PBR^{-1}B^T P + Q < 0, \tag{2.6}$$

where A, B, $Q = Q^T$, $R = R^T > 0$ are given matrices of appropriate sizes, and $P = P^T$ is the variable. Note that this is a *quadratic* matrix inequality in the variable P. It can be expressed as the *linear* matrix inequality

$$\begin{bmatrix} -A^TP - PA - Q & PB \\ B^TP & R \end{bmatrix} > 0.$$

This representation also clearly shows that the quadratic matrix inequality (2.6) is convex in P, which is not obvious.

2.1.2 Linear equality constraints

In some problems we will encounter linear equality constraints on the variables, e.g.

$$P > 0, \qquad A^TP + PA < 0, \qquad \mathbf{Tr}\, P = 1, \tag{2.7}$$

where $P \in \mathbf{R}^{k\times k}$ is the variable. Of course we can eliminate the equality constraint to write (2.7) in the form $F(x) > 0$. Let $P_1, \ldots, P_m$ be a basis for symmetric $k\times k$ matrices with trace zero ($m = (k(k+1)/2) - 1$) and let P_0 be a symmetric $k \times k$ matrix with $\mathbf{Tr}\, P_0 = 1$. Then take $F_0 = \mathbf{diag}(P_0, -A^TP_0 - P_0A)$ and $F_i = \mathbf{diag}(P_i, -A^TP_i - P_iA)$ for $i = 1, \ldots, m$.

We will refer to constraints such as (2.7) as LMIs, leaving any required elimination of equality constraints to the reader.

2.2 Some Standard Problems

Here we list some common convex and quasiconvex problems that we will encounter in the sequel.

2.2.1 LMI problems

Given an LMI $F(x) > 0$, the corresponding LMI Problem (LMIP) is to find x^{feas} such that $F(x^{\mathrm{feas}}) > 0$ or determine that the LMI is infeasible. (By duality, this means: Find a nonzero $G \geq 0$ such that $\mathbf{Tr}\, GF_i = 0$ for $i = 1, \ldots, m$ and $\mathbf{Tr}\, GF_0 \leq 0$; see the Notes and References.) Of course, this is a convex feasibility problem. We will say "solving the LMI $F(x) > 0$" to mean solving the corresponding LMIP.

As an example of an LMIP, consider the "simultaneous Lyapunov stability problem" (which we will see in §5.1): We are given $A_i \in \mathbf{R}^{n\times n}$, $i = 1, \ldots, L$, and need to find P satisfying the LMI

$$P > 0, \qquad A_i^TP + PA_i < 0, \qquad i = 1, \ldots, L,$$

or determine that no such P exists. Determining that no such P exists is equivalent to finding $Q_0, \ldots, Q_L$, not all zero, such that

$$Q_0 \geq 0, \ldots, Q_L \geq 0, \qquad Q_0 = \sum_{i=1}^{L} \left(Q_iA_i^T + A_iQ_i\right), \tag{2.8}$$

which is another (nonstrict) LMIP.

2.2.2 Eigenvalue problems

The eigenvalue problem (EVP) is to minimize the maximum eigenvalue of a matrix that depends affinely on a variable, subject to an LMI constraint (or determine that the constraint is infeasible), i.e.,

$$\begin{array}{ll} \text{minimize} & \lambda \\ \text{subject to} & \lambda I - A(x) > 0, \qquad B(x) > 0 \end{array}$$

where A and B are symmetric matrices that depend affinely on the optimization variable x. This is a convex optimization problem.

EVPs can appear in the equivalent form of minimizing a linear function subject to an LMI, i.e.,

$$\begin{array}{ll} \text{minimize} & c^T x \\ \text{subject to} & F(x) > 0 \end{array} \tag{2.9}$$

with F an affine function of x. In the special case when the matrices F_i are all diagonal, this problem reduces to the general linear programming problem: minimizing the linear function $c^T x$ subject to a set of linear inequalities on x.

Another equivalent form for the EVP is:

$$\begin{array}{ll} \text{minimize} & \lambda \\ \text{subject to} & A(x, \lambda) > 0 \end{array}$$

where A is affine in (x, λ). We leave it to the reader to verify that these forms are equivalent, i.e., any can be transformed into any other.

As an example of an EVP, consider the problem (which appears in §6.3.2):

$$\begin{array}{ll} \text{minimize} & \gamma \\ \text{subject to} & \begin{bmatrix} -A^T P - PA - C^T C & PB \\ B^T P & \gamma I \end{bmatrix} > 0, \qquad P > 0 \end{array}$$

where the matrices $A \in \mathbf{R}^{n\times n}$, $B \in \mathbf{R}^{n\times p}$, and $C \in \mathbf{R}^{m\times n}$ are given, and P and γ are the optimization variables. From our remarks above, this EVP can also be expressed in terms of the associated quadratic matrix inequality:

$$\begin{array}{ll} \text{minimize} & \gamma \\ \text{subject to} & A^T P + PA + C^T C + \gamma^{-1} PBB^T P < 0, \qquad P > 0 \end{array}$$

2.2.3 Generalized eigenvalue problems

The generalized eigenvalue problem (GEVP) is to minimize the maximum generalized eigenvalue of a pair of matrices that depend affinely on a variable, subject to an LMI constraint. The general form of a GEVP is:

$$\begin{array}{ll} \text{minimize} & \lambda \\ \text{subject to} & \lambda B(x) - A(x) > 0, \qquad B(x) > 0, \qquad C(x) > 0 \end{array}$$

where A, B and C are symmetric matrices that are affine functions of x. We can express this as

$$\begin{array}{ll} \text{minimize} & \lambda_{\max}(A(x), B(x)) \\ \text{subject to} & B(x) > 0, \qquad C(x) > 0 \end{array}$$

where $\lambda_{\max}(X,Y)$ denotes the largest generalized eigenvalue of the pencil $\lambda Y - X$ with $Y > 0$, i.e., the largest eigenvalue of the matrix $Y^{-1/2}XY^{-1/2}$. This GEVP is a quasiconvex optimization problem since the constraint is convex and the objective, $\lambda_{\max}(A(x),B(x))$, is quasiconvex. This means that for feasible x, $\tilde{x}$ and $0 \le \theta \le 1$,

$$\lambda_{\max}(A(\theta x + (1-\theta)\tilde{x}), B(\theta x + (1-\theta)\tilde{x})) \le \\ \max\left\{\lambda_{\max}(A(x),B(x)), \lambda_{\max}(A(\tilde{x}),B(\tilde{x}))\right\}.$$

Note that when the matrices are all diagonal and $A(x)$ and $B(x)$ are scalar, this problem reduces to the general linear-fractional programming problem, i.e., minimizing a linear-fractional function subject to a set of linear inequalities. In addition, many nonlinear quasiconvex functions can be represented in the form of a GEVP with appropriate A, B, and C; see the Notes and References.

An equivalent alternate form for a GEVP is

$$\begin{array}{ll} \text{minimize} & \lambda \\ \text{subject to} & A(x,\lambda) > 0 \end{array}$$

where $A(x,\lambda)$ is affine in x for fixed λ and affine in λ for fixed x, and satisfies the monotonicity condition $\lambda > \mu \implies A(x,\lambda) \ge A(x,\mu)$. As an example of a GEVP, consider the problem

$$\begin{array}{ll} \text{maximize} & \alpha \\ \text{subject to} & -A^TP - PA - 2\alpha P > 0, \qquad P > 0 \end{array}$$

where the matrix A is given, and the optimization variables are the symmetric matrix P and the scalar α. (This problem arises in §5.1.)

2.2.4 A convex problem

Although we will be concerned mostly with LMIPs, EVPs, and GEVPs, we will also encounter the following convex problem, which we will abbreviate CP:

$$\begin{array}{ll} \text{minimize} & \log\det A(x)^{-1} \\ \text{subject to} & A(x) > 0, \qquad B(x) > 0 \end{array} \tag{2.10}$$

where A and B are symmetric matrices that depend affinely on x. (Note that when $A > 0$, $\log\det A^{-1}$ is a convex function of A.)

Remark: Problem CP can be transformed into an EVP, since $\det A(x) > \lambda$ can be represented as an LMI in x and λ; see [NN94, §6.4.3]. As we do with variables which are matrices, we leave these problems in the more natural form.

As an example of CP, consider the problem:

$$\begin{array}{ll} \text{minimize} & \log\det P^{-1} \\ \text{subject to} & P > 0, \qquad v_i^TPv_i \le 1, \quad i = 1,\ldots,L \end{array} \tag{2.11}$$

here $v_i \in \mathbf{R}^n$ are given and $P = P^T \in \mathbf{R}^{n\times n}$ is the variable.

We will encounter several variations of this problem, which has the following interpretation. Let $\mathcal{E}$ denote the ellipsoid centered at the origin determined by P,

$\mathcal{E} \triangleq \{z \mid z^T P z \leq 1\}$. The constraints are simply $v_i \in \mathcal{E}$. Since the volume of $\mathcal{E}$ is proportional to $(\det P)^{-1/2}$, minimizing $\log\det P^{-1}$ is the same as minimizing the volume of $\mathcal{E}$. So by solving (2.11), we find the minimum volume ellipsoid, centered at the origin, that contains the points $v_1, \ldots, v_L$, or equivalently, the polytope $\mathbf{Co}\{v_1, \ldots, v_L\}$, where $\mathbf{Co}$ denotes the convex hull.

2.2.5 Solving these problems

The standard problems (LMIPs, EVPs, GEVPs, and CPs) are *tractable*, from both theoretical and practical viewpoints:

- They can be solved in polynomial-time (indeed with a variety of interpretations for the term "polynomial-time").
- They can be solved in practice very efficiently.

By "solve the problem" we mean: Determine whether or not the problem is feasible, and if it is, compute a feasible point with an objective value that exceeds the global minimum by less than some prespecified accuracy.

2.3 Ellipsoid Algorithm

We first describe a simple ellipsoid algorithm that, roughly speaking, is guaranteed to solve the standard problems. We describe it here because it is very simple and, from a theoretical point of view, efficient (polynomial-time). In practice, however, the interior-point algorithms described in the next section are much more efficient.

Although more sophisticated versions of the ellipsoid algorithm can detect infeasible constraints, we will assume that the problem we are solving has at least one optimal point, i.e., the constraints are feasible. (In the feasibility problem, we consider any feasible point as being optimal.) The basic idea of the algorithm is as follows. We start with an ellipsoid $\mathcal{E}^{(0)}$ that is guaranteed to contain an optimal point. We then compute a *cutting plane* for our problem that passes through the center $x^{(0)}$ of $\mathcal{E}^{(0)}$. This means that we find a nonzero vector $g^{(0)}$ such that an optimal point lies in the half-space $\{z \mid g^{(0)T}(z - x^{(0)}) \leq 0\}$ (or in the half-space $\{z \mid g^{(0)T}(z - x^{(0)}) < 0\}$, depending on the situation). (We will explain how to do this for each of our problems later.) We then know that the sliced half-ellipsoid

$$\mathcal{E}^{(0)} \cap \left\{z \,\middle|\, g^{(0)T}(z - x^{(0)}) \leq 0\right\}$$

contains an optimal point. Now we compute the ellipsoid $\mathcal{E}^{(1)}$ of minimum volume that contains this sliced half-ellipsoid; $\mathcal{E}^{(1)}$ is then guaranteed to contain an optimal point. The process is then repeated.

We now describe the algorithm more explicitly. An ellipsoid $\mathcal{E}$ can be described as

$$\mathcal{E} = \{z \mid (z-a)^T A^{-1} (z-a) \leq 1\}$$

where $A = A^T > 0$. The minimum volume ellipsoid that contains the half-ellipsoid

$$\{z \mid (z-a)^T A^{-1}(z-a) \leq 1, \quad g^T(z-a) \leq 0\}$$

is given by

$$\tilde{\mathcal{E}} = \left\{z \,\middle|\, (z-\tilde{a})^T \tilde{A}^{-1}(z-\tilde{a}) \leq 1\right\},$$

where

$$\tilde{a} = a - \frac{A\tilde{g}}{m+1}, \qquad \tilde{A} = \frac{m^2}{m^2-1}\left(A - \frac{2}{m+1}A\tilde{g}\tilde{g}^T A\right),$$

and $\tilde{g} = g/\sqrt{g^T A g}$. (We note that these formulas hold only for $m \geq 2$. In the case of one variable, the minimum length interval containing a half-interval is the half-interval itself; the ellipsoid algorithm, in this case, reduces to the familiar bisection algorithm.)

The ellipsoid algorithm is initialized with $x^{(0)}$ and $A^{(0)}$ such that the corresponding ellipsoid contains an optimal point. The algorithm then proceeds as follows: for $k = 1, 2, \ldots$

compute a $g^{(k)}$ that defines a cutting plane at $x^{(k)}$

$$\tilde{g} := \left(g^{(k)T} A^{(k)} g^{(k)}\right)^{-1/2} g^{(k)}$$

$$x^{(k+1)} := x^{(k)} - \tfrac{1}{m+1} A^{(k)} \tilde{g}$$

$$A^{(k+1)} := \tfrac{m^2}{m^2-1}\left(A^{(k)} - \tfrac{2}{m+1} A^{(k)} \tilde{g}\tilde{g}^T A^{(k)}\right)$$

This recursion generates a sequence of ellipsoids that are guaranteed to contain an optimal point. It turns out that the volume of these ellipsoids decreases geometrically. We have

$$\mathbf{vol}(\mathcal{E}^{(k)}) \leq e^{-\frac{k}{2m}}\, \mathbf{vol}(\mathcal{E}^{(0)}),$$

and this fact can be used to prove polynomial-time convergence of the algorithm. (We refer the reader to the papers cited in the Notes and References for precise statements of what we mean by "polynomial-time" and indeed, "convergence," as well as proofs.)

We now show how to compute cutting planes for each of our standard problems.

LMIPs: Consider the LMI

$$F(x) = F_0 + \sum_{i=1}^m x_i F_i > 0.$$

If x does not satisfy this LMI, there exists a nonzero u such that

$$u^T F(x) u = u^T \left(F_0 + \sum_{i=1}^m x_i F_i\right) u \leq 0.$$

Define g by $g_i = -u^T F_i u$, $i = 1, \ldots, m$. Then for any z satisfying $g^T(z - x) \geq 0$ we have

$$u^T F(z) u = u^T \left(F_0 + \sum_{i=1}^m z_i F_i\right) u = u^T F(x) u - g^T(z-x) \leq 0.$$

It follows that every feasible point lies in the half-space $\{z \mid g^T(z-x) < 0\}$, i.e., this g defines a cutting plane for the LMIP at the point x.

EVPs: Consider the EVP

$$\begin{array}{ll} \text{minimize} & c^T x \\ \text{subject to} & F(x) > 0 \end{array}$$

Suppose first that the given point x is infeasible, i.e., $F(x) \not> 0$. Then we can construct a cutting plane for this problem using the method described above for LMIPs. In this case, we are discarding the half-space $\{z \mid g^T(z-x) \geq 0\}$ because all such points are infeasible.

Now assume that the given point x is feasible, i.e., $F(x) > 0$. In this case, the vector $g = c$ defines a cutting plane for the EVP at the point x. Here, we are discarding the half-space $\{z \mid g^T(z-x) > 0\}$ because all such points (whether feasible or not) have an objective value larger than x, and hence cannot be optimal.

GEVPs: We consider the formulation

$$\begin{array}{ll} \text{minimize} & \lambda_{\max}(A(x), B(x)) \\ \text{subject to} & B(x) > 0, \qquad C(x) > 0 \end{array}$$

Here, $A(x) = A_0 + \sum_{i=1}^m x_i A_i$, $B(x) = B_0 + \sum_{i=1}^m x_i B_i$ and $C(x) = C_0 + \sum_{i=1}^m x_i C_i$. Now suppose we are given a point x. If the constraints are violated, we use the method described for LMIPs to generate a cutting plane at x.

Suppose now that x is feasible. Pick any $u \neq 0$ such that

$$(\lambda_{\max}(A(x), B(x))B(x) - A(x))\, u = 0.$$

Define g by

$$g_i = -u^T(\lambda_{\max}(A(x), B(x))B_i - A_i)u, \qquad i = 1, \ldots, m.$$

We claim g defines a cutting plane for this GEVP at the point x. To see this, we note that for any z,

$$u^T\left(\lambda_{\max}(A(x), B(x))B(z) - A(z)\right)u = -g^T(z-x).$$

Hence, if $g^T(z-x) \geq 0$ we find that

$$\lambda_{\max}(A(z), B(z)) \geq \lambda_{\max}(A(x), B(x))$$

which establishes our claim.

CPs: We now consider our standard CP (2.10). When the given point x is infeasible we already know how to generate a cutting plane. So we assume that x is feasible. In this case, a cutting plane is given by the gradient of the objective

$$-\log\det A(x) = -\log\det\left(A_0 + \sum_{i=1}^m x_i A_i\right)$$

at x, i.e., $g_i = -\,\mathbf{Tr}\, A_i A(x)^{-1}$. Since the objective function is convex we have for all z,

$$\log\det A(z)^{-1} \geq \log\det A(x)^{-1} + g^T(z-x).$$

In particular, $g^T(z-x) > 0$ implies $\log\det A(z)^{-1} > \log\det A(x)^{-1}$, and hence z cannot be optimal.

2.4 Interior-Point Methods

Since 1988, interior-point methods have been developed for the standard problems. In this section we describe a simple interior-point method for solving an EVP, given a feasible starting point. The references describe more sophisticated interior-point methods for the other problems (including the problem of computing a feasible starting point for an EVP).

2.4.1 Analytic center of an LMI

The notion of the analytic center of an LMI plays an important role in interior-point methods, and is important in its own right. Consider the LMI

$$F(x) = F_0 + \sum_{i=1}^{m} x_i F_i > 0,$$

where $F_i = F_i^T \in \mathbf{R}^{n\times n}$. The function

$$\phi(x) \triangleq \begin{cases} \log\det F(x)^{-1} & F(x) > 0 \\ \infty & \text{otherwise,} \end{cases} \tag{2.12}$$

is finite if and only if $F(x) > 0$, and becomes infinite as x approaches the boundary of the feasible set $\{x | F(x) > 0\}$, i.e., it is a *barrier function* for the feasible set. (We have already encountered this function in our standard problem CP.)

We suppose now that the feasible set is nonempty and bounded, which implies that the matrices $F_1, \ldots, F_m$ are linearly independent (otherwise the feasible set will contain a line). It can be shown that ϕ is strictly convex on the feasible set, so it has a unique minimizer, which we denote $x^\star$:

$$x^\star \triangleq \arg\min_x \ \phi(x). \tag{2.13}$$

We refer to $x^\star$ as the *analytic center* of the LMI $F(x) > 0$. Equivalently,

$$x^\star = \arg\max_{F(x)>0} \ \det F(x), \tag{2.14}$$

that is, $F(x^\star)$ has maximum determinant among all positive-definite matrices of the form $F(x)$.

> **Remark**: The analytic center depends on the LMI (i.e., the data $F_0, \ldots, F_m$) and not its feasible set: We can have two LMIs with different analytic centers but identical feasible sets. The analytic center is, however, invariant under congruence of the LMI: $F(x) > 0$ and $T^T F(x) T > 0$ have the same analytic center provided T is nonsingular.

We now turn to the problem of computing the analytic center of an LMI. (This is a special form of our problem CP.) Newton's method, with appropriate step length selection, can be used to efficiently compute $x^\star$, starting from a feasible initial point. We consider the algorithm:

$$x^{(k+1)} := x^{(k)} - \alpha^{(k)} H(x^{(k)})^{-1} g(x^{(k)}), \tag{2.15}$$

where $\alpha^{(k)}$ is the damping factor of the kth iteration, and $g(x^{(k)})$ and $H(x^{(k)})$ denote the gradient and Hessian of ϕ, respectively, at $x^{(k)}$. In [NN94, §2.2], Nesterov and Nemirovskii give a simple step length rule appropriate for a general class of barrier functions (called *self-concordant*), and in particular, the function ϕ. Their damping factor is:

$$\alpha^{(k)} := \begin{cases} 1 & \text{if } \delta(x^{(k)}) \le 1/4, \\ 1/(1+\delta(x^{(k)})) & \text{otherwise,} \end{cases} \tag{2.16}$$

where

$$\delta(x) \stackrel{\Delta}{=} \sqrt{g(x)^T H(x)^{-1} g(x)}$$

is called the *Newton decrement* of ϕ at x. Nesterov and Nemirovskii show that this step length always results in $x^{(k+1)}$ feasible, i.e., $F(x^{(k+1)}) > 0$, and convergence of $x^{(k)}$ to $x^\star$.

Indeed, they give sharp bounds on the number of steps required to compute $x^\star$ to a given accuracy using Newton's method with the step length (2.16). They prove that $\phi(x^{(k)}) - \phi(x^\star) \leq \epsilon$ whenever

$$k \geq c_1 + c_2 \log\log(1/\epsilon) + c_3 \left(\phi(x^{(0)}) - \phi(x^\star)\right), \tag{2.17}$$

where c_1, c_2, and c_3 are three absolute constants, i.e., specific numbers. The first and second terms on the right-hand side do not depend on any problem data, i.e., the matrices $F_0, \ldots, F_m$, and the numbers m and n. The second term grows so slowly with required accuracy ϵ that for all practical purposes it can be lumped together with the first and considered an absolute constant. The last term on the right-hand side of (2.17) depends on how "centered" the initial point is.

2.4.2 The path of centers

Now consider the standard EVP:

$$\begin{array}{ll} \text{minimize} & c^T x \\ \text{subject to} & F(x) > 0 \end{array}$$

Let λ^{opt} denote its optimal value, so for each $\lambda > \lambda^{\text{opt}}$ the LMI

$$F(x) > 0, \qquad c^T x < \lambda \tag{2.18}$$

is feasible. We will also assume that the LMI (2.18) has a bounded feasible set, and therefore has an analytic center which we denote $x^\star(\lambda)$:

$$x^\star(\lambda) \stackrel{\Delta}{=} \underset{x}{\text{arg min}} \left(\log\det F(x)^{-1} + \log \frac{1}{\lambda - c^T x}\right).$$

The curve given by $x^\star(\lambda)$ for $\lambda > \lambda^{\text{opt}}$ is called the *path of centers* for the EVP. It can be shown that it is analytic and has a limit as $\lambda \to \lambda^{\text{opt}}$, which we denote x^{opt}. The point x^{opt} is optimal (or more precisely, the limit of a minimizing sequence) since for $\lambda > \lambda^{\text{opt}}$, $x^\star(\lambda)$ is feasible and satisfies $c^T x^\star(\lambda) < \lambda$.

The point $x^\star(\lambda)$ is characterized by

$$\begin{aligned} &\left.\frac{\partial}{\partial x_i}\right|_{x^\star(\lambda)} \left(\log\det F(x)^{-1} + \log \frac{1}{\lambda - c^T x}\right) \\ &= -\,\mathbf{Tr}\, F(x^\star(\lambda))^{-1} F_i + \frac{c_i}{\lambda - c^T x^\star(\lambda)} = 0, \quad i = 1, \ldots, m. \end{aligned} \tag{2.19}$$

2.4.3 Method of centers

The method of centers is a simple interior-point algorithm that solves an EVP, given a feasible starting point. The algorithm is initialized with $\lambda^{(0)}$ and $x^{(0)}$, with $F(x^{(0)}) > 0$ and $c^T x^{(0)} < \lambda^{(0)}$, and proceeds as follows:

$$\begin{aligned} \lambda^{(k+1)} &:= (1-\theta) c^T x^{(k)} + \theta \lambda^{(k)} \\ x^{(k+1)} &:= x^\star(\lambda^{(k+1)}) \end{aligned}$$

where θ is a parameter with $0 < \theta < 1$.

The classic method of centers is obtained with $\theta = 0$. In this case, however, $x^{(k)}$ does not (quite) satisfy the new inequality $c^T x < \lambda^{(k+1)}$. With $\theta > 0$, however, the current iterate $x^{(k)}$ is feasible for the inequality $c^T x < \lambda^{(k+1)}$, $F(x) > 0$, and therefore can be used as the initial point for computing the next iterate $x^\star(\lambda^{(k+1)})$ via Newton's method. Since computing an analytic center is itself a special type of CP, we can view the method of centers as a way of solving an EVP by solving a sequence of CPs (which is done using Newton's method).

We now give a simple proof of convergence. Multiplying (2.19) by $(x_i^{(k)} - x_i^{\rm opt})$ and summing over i yields:

$$\mathbf{Tr}\, F(x^{(k)})^{-1}\left(F(x^{(k)}) - F(x^{\rm opt})\right) = \frac{1}{\lambda^{(k)} - c^T x^{(k)}} c^T (x^{(k)} - x^{\rm opt}).$$

Since $\mathbf{Tr}\, F(x^{(k)})^{-1} F(x^{\rm opt}) \geq 0$, we conclude that

$$n \geq \frac{1}{\lambda^{(k)} - c^T x^{(k)}} (c^T x^{(k)} - \lambda^{\rm opt}).$$

Replacing $c^T x^{(k)}$ by $(\lambda^{(k+1)} - \theta\lambda^{(k)})/(1-\theta)$ yields

$$(\lambda^{(k+1)} - \lambda^{\rm opt}) \leq \frac{n+\theta}{n+1}(\lambda^{(k)} - \lambda^{\rm opt}),$$

which proves that $\lambda^{(k)}$ approaches $\lambda^{\rm opt}$ with at least geometric convergence. Note that we can also express the inequality above in the form

$$c^T x^{(k)} - \lambda^{\rm opt} \leq n(\lambda^{(k)} - c^T x^{(k)}),$$

which shows that the stopping criterion

$$\text{until } \left(\lambda^{(k)} - c^T x^{(k)} < \epsilon/n\right)$$

guarantees that on exit, the optimal value has been found within ϵ.

We make a few comments here, and refer the reader to the Notes and References for further elaboration. First, this variation on the method of centers is not polynomial-time, but more sophisticated versions are. Second, and perhaps more important, we note that two simple modifications of the method of centers as described above yield an algorithm that is fairly efficient in practice. The modifications are:

- Instead of the Nesterov–Nemirovskii step length, a step length chosen to minimize ϕ along the Newton search direction (i.e., an exact line-search) will yield faster convergence to the analytic center.
- Instead of defining $x^\star(\lambda)$ to be the analytic center of the LMI $F(x) > 0$, $c^T x < \lambda$, we define it to be the analytic center of the LMI $F(x) > 0$ along with q copies of $c^T x < \lambda$, where $q > 1$ is an integer. In other words we use

$$x^\star(\lambda) \stackrel{\Delta}{=} \mathop{\rm arg\ min}_{x} \left(\log\det F(x)^{-1} + q \log \frac{1}{\lambda - c^T x}\right).$$

  Using $q > 1$, say $q \approx n$ where $F(x) \in \mathbf{R}^{n\times n}$, yields much faster reduction of λ per iteration.

Among interior-point methods for the standard problems, the method of centers is not the most efficient. The most efficient algorithms developed so far appear to be primal-dual algorithms (and variations) and the projective method of Nemirovskii; see the Notes and References.

2.4.4 Interior-point methods and problem structure

An important feature of interior-point methods is that problem structure can be exploited to increase efficiency. The idea is very roughly as follows. In interior-point methods most of the computational effort is devoted to computing the Newton direction of a barrier or similar function. It turns out that this Newton direction can be expressed as the solution of a weighted least-squares problem of the same size as the original problem. Using conjugate-gradient and other related methods to solve these least-squares systems gives two advantages. First, by exploiting problem structure in the conjugate-gradient iterations, the computational effort required to solve the least-squares problems is much smaller than by standard "direct" methods such as QR or Cholesky factorization. Second, it is possible to terminate the conjugate-gradient iterations before convergence, and still obtain an approximation of the Newton direction suitable for interior-point methods. See the Notes and References for more discussion.

An example will demonstrate the efficiencies that can be obtained using the techniques sketched above. The problem is an EVP that we will encounter in §6.2.1.

We are given matrices $A_1, \ldots, A_L \in \mathbf{R}^{n \times n}$, symmetric matrices $D_1, \ldots, D_L, E \in \mathbf{R}^{n \times n}$. We consider the EVP

$$\begin{array}{ll} \text{minimize} & \mathbf{Tr}\, EP \\ \text{subject to} & A_i^T P + P A_i + D_i < 0, \quad i = 1, \ldots, L \end{array} \tag{2.20}$$

In this problem the variable is the matrix P, so the dimension of the optimization variable is $m = n(n+1)/2$. When the Lyapunov inequalities are combined into one large LMI $F(x) > 0$, we find that $F(x) \in \mathbf{R}^{N \times N}$ with $N = Ln$. This LMI has much structure: It is block-diagonal with each block a Lyapunov inequality.

Vandenberghe and Boyd have developed a (primal-dual) interior-point method that solves (2.20), exploiting the problem structure. They *prove* the worst-case estimate of $O(m^{2.75} L^{1.5})$ arithmetic operations to solve the problem to a given accuracy. In comparison, the ellipsoid method solves (2.20) to a given accuracy in $O(m^{3.5} L)$ arithmetic operations (moreover, the constant hidden in the $O(\cdot)$ notation is much larger for the ellipsoid algorithm).

Numerical experiments on families of problems with randomly generated data and families of problems arising in system and control theory show that the actual performance of the interior-point method is much better than the worst-case estimate: $O(m^\alpha L^\beta)$ arithmetic operations, with $\alpha \approx 2.1$ and $\beta \approx 1.2$. This complexity estimate is remarkably small. For example, the cost of solving L Lyapunov equations of the same size is $O(m^{1.5} L)$. Therefore, the relative cost of solving L coupled Lyapunov inequalities, compared to the cost of solving L independent Lyapunov inequalities is $O(m^{0.6} L^{0.2})$. This example illustrates one of our points: The computational cost of solving one of the standard problems (which has no "analytic solution") can be comparable to the computational cost of evaluating the solution of a similar problem that has an "analytic solution".

2.5 Strict and Nonstrict LMIs

We have so far assumed that the optimization problems LMIP, EVP, GEVP, and CP involve strict LMIs. We will also encounter these problems with nonstrict LMIs, or more generally, with a mixture of strict and nonstrict LMIs.

As an example consider the nonstrict version of the EVP (2.9), i.e.

$$\begin{array}{ll} \text{minimize} & c^T x \\ \text{subject to} & F(x) \geq 0 \end{array}$$

Intuition suggests that we could simply solve the strict EVP (by, say, an interior-point method) to obtain the solution of the nonstrict EVP. This is correct in most but not all cases.

If the strict LMI $F(x) > 0$ is feasible, then we have

$$\{x \in \mathbf{R}^m \mid F(x) \geq 0\} = \overline{\{x \in \mathbf{R}^m \mid F(x) > 0\}}, \tag{2.21}$$

i.e., the feasible set of the nonstrict LMI is the closure of the feasible set of the strict LMI. It follows that

$$\inf\left\{c^T x \mid F(x) \geq 0\right\} = \inf\left\{c^T x \mid F(x) > 0\right\}. \tag{2.22}$$

So in this case, we can solve the strict EVP to obtain a solution of the nonstrict EVP. This is true for the problems GEVP and CP as well.

We will say that the LMI $F(x) \geq 0$ is *strictly feasible* if its strict version is feasible, i.e., if there is some $x_0 \in \mathbf{R}^m$ such that $F(x_0) > 0$. We have just seen that when an LMI is strictly feasible, we can replace nonstrict inequality with strict inequality in the problems EVP, GEVP, and CP in order to solve them. In the language of optimization theory, the requirement of strict feasibility is a (very strong) constraint qualification.

When an LMI is feasible but not strictly feasible, (2.21) need not hold, and the EVPs with the strict and nonstrict LMIs can be very different. As a simple example, consider $F(x) = \mathbf{diag}(x, -x)$ with $x \in \mathbf{R}$. The right-hand side of (2.22) is $+\infty$ since the strict LMI $F(x) > 0$ is infeasible. The left-hand side, however, is always 0, since the LMI $F(x) \geq 0$ has the single feasible point $x = 0$. This example shows one of the two pathologies that can occur: The nonstrict inequality contains an implicit equality constraint (in contrast with an explicit equality constraint as in §2.1.2).

The other pathology is demonstrated by the example $F(x) = \mathbf{diag}(x, 0)$ with $x \in \mathbf{R}$ and $c = -1$. Once again, the strict LMI is infeasible so the right-hand side of (2.22) is $+\infty$. The feasible set for the nonstrict LMI is the interval $[0, \infty)$ so the right-hand side is $-\infty$. The problem here is that $F(x)$ is always singular. Of course, the nonstrict LMI $F(x) \geq 0$ is equivalent (in the sense of defining equal feasible sets) to the "reduced" LMI $\tilde{F}(x) = x \geq 0$. Note that this reduced LMI satisfies the constraint qualification, i.e., is strictly feasible.

2.5.1 Reduction to a strictly feasible LMI

It turns out that any feasible nonstrict LMI can be reduced to an equivalent LMI that is strictly feasible, by eliminating implicit equality constraints and then reducing the resulting LMI by removing any constant nullspace.

The precise statement is: Let $F_0, \ldots, F_m \in \mathbf{R}^{n \times n}$ be symmetric. Then there is a matrix $A \in \mathbf{R}^{m \times p}$ with $p \leq m$, a vector $b \in \mathbf{R}^m$, and symmetric matrices $\tilde{F}_0, \ldots, \tilde{F}_p \in \mathbf{R}^{q \times q}$ with $q \leq n$ such that:

$$F(x) \geq 0 \quad \Longleftrightarrow \quad \begin{array}{l} x = Az + b \text{ for some } z \in \mathbf{R}^p \text{ with} \\ \tilde{F}(z) \triangleq \tilde{F}_0 + \sum_{i=1}^{p} z_i \tilde{F}_i \geq 0 \end{array}$$

where the LMI $\tilde{F}(z) \geq 0$ is either infeasible or strictly feasible. See the Notes and References for a proof.

The matrix A and vector b describe the implicit equality constraints for the LMI $F(x) \geq 0$. Similarly, the LMI $\tilde{F}(z) \geq 0$ can be interpreted as the original LMI with its constant nullspace removed (see the Notes and References). In most of the problems encountered in this book, there are no implicit equality constraints or nontrivial common nullspace for F, so we can just take $A = I$, $b = 0$, and $\tilde{F} = F$.

Using this reduction we can, at least in principle, always deal with strictly feasible LMIs. For example we have

$$\begin{aligned} \inf \left\{ c^T x \mid F(x) \geq 0 \right\} &= \inf \left\{ c^T (Az + b) \;\middle|\; \tilde{F}(z) \geq 0 \right\} \\ &= \inf \left\{ c^T (Az + b) \;\middle|\; \tilde{F}(z) > 0 \right\} \end{aligned}$$

since the LMI $\tilde{F}(z) \geq 0$ is either infeasible or strictly feasible.

2.5.2 Example: Lyapunov inequality

To illustrate the previous ideas we consider the simplest LMI arising in control theory, the Lyapunov inequality:

$$A^T P + PA \leq 0, \qquad P > 0, \tag{2.23}$$

where $A \in \mathbf{R}^{k \times k}$ is given, and the symmetric matrix P is the variable. Note that this LMI contains a strict inequality as well as a nonstrict inequality.

We know from system theory that the LMI (2.23) is feasible if and only if all trajectories of $\dot{x} = Ax$ are bounded, or equivalently, if the eigenvalues of A have nonpositive real part, and those with zero real part are nondefective, i.e., correspond to Jordan blocks of size one. We also know from system theory that the LMI (2.23) is strictly feasible if and only if all trajectories of $\dot{x} = Ax$ converge to zero, or equivalently, all the eigenvalues of A have negative real part.

Consider the interesting case where the LMI (2.23) is feasible but not strictly feasible. From the remarks above, we see that by a change of coordinates we can put A in the form

$$\begin{aligned} \tilde{A} &\stackrel{\Delta}{=} T^{-1} A T \\ &= \mathbf{diag}\left(\begin{bmatrix} 0 & \omega_1 I_{k_1} \\ -\omega_1 I_{k_1} & 0 \end{bmatrix}, \ldots, \begin{bmatrix} 0 & \omega_r I_{k_r} \\ -\omega_r I_{k_r} & 0 \end{bmatrix}, 0_{k_{r+1}}, A_{\rm stab} \right) \end{aligned}$$

where $0 < \omega_1 < \cdots < \omega_r$, $0_{k_{r+1}}$ denotes the zero matrix in $\mathbf{R}^{k_{r+1} \times k_{r+1}}$, and all the eigenvalues of the matrix $A_{\rm stab} \in \mathbf{R}^{s \times s}$ have negative real part. Roughly speaking, we have separated out the stable part of A, the part corresponding to each imaginary axis eigenvalue, and the part associated with eigenvalue zero.

Using standard methods of Lyapunov theory it can be shown that

$$\{P \mid A^TP + PA \le 0,\ P > 0\} =$$

$$\left\{ T^{-T}\tilde{P}T^{-1} \;\middle|\; \begin{array}{l} \tilde{P} = \mathbf{diag}\left(\begin{bmatrix} P_1 & Q_1 \\ Q_1^T & P_1 \end{bmatrix}, \ldots, \begin{bmatrix} P_r & Q_r \\ Q_r^T & P_r \end{bmatrix}, P_{r+1}, P_{\rm stab} \right) \\ P_i \in \mathbf{R}^{k_i \times k_i},\ i = 1, \ldots, r+1, \quad Q_i^T = -Q_i,\ i = 1, \ldots, r \\ \begin{bmatrix} P_i & Q_i \\ Q_i^T & P_i \end{bmatrix} > 0,\ i = 1, \ldots, r, \quad P_{r+1} > 0,\ P_{\rm stab} > 0 \\ A_{\rm stab}^T P_{\rm stab} + P_{\rm stab} A_{\rm stab} \le 0 \end{array} \right\}.$$

From this characterization we can find a reduced LMI that is strictly feasible. We can take the symmetric matrices $P_1, \ldots, P_{r+1}, P_{\rm stab}$ and the skew-symmetric matrices $Q_1, \ldots, Q_r$ as the "free" variable z; the affine mapping from z into x simply maps these matrices into

$$P = T^{-T}\,\mathbf{diag}\left(\begin{bmatrix} P_1 & Q_1 \\ Q_1^T & P_1 \end{bmatrix}, \ldots, \begin{bmatrix} P_r & Q_r \\ Q_r^T & P_r \end{bmatrix}, P_{r+1}, P_{\rm stab} \right) T^{-1}. \qquad (2.24)$$

Put another way, the equality constraints implicit in the LMI (2.23) are that P must have this special structure.

Now we substitute P in the form (2.24) back into the original LMI (2.23). We find that

$$A^TP + PA = T^T\,\mathbf{diag}\left(0, A_{\rm stab}^T P_{\rm stab} + P_{\rm stab} A_{\rm stab}\right) T$$

where the zero matrix has size $2k_1 + \cdots + 2k_r + k_{r+1}$. We remove the constant nullspace to obtain the reduced version, i.e., $A_{\rm stab}^T P_{\rm stab} + P_{\rm stab} A_{\rm stab} \le 0$. Thus, the reduced LMI corresponding to (2.23) is

$$\begin{array}{l} \begin{bmatrix} P_i & Q_i \\ Q_i^T & P_i \end{bmatrix} > 0,\ i = 1, \ldots, r, \quad P_{r+1} > 0, \\ P_{\rm stab} > 0,\ A_{\rm stab}^T P_{\rm stab} + P_{\rm stab} A_{\rm stab} \le 0. \end{array} \qquad (2.25)$$

This reduced LMI is strictly feasible, since we can take $P_1, \ldots, P_{r+1}$ as identity matrices, $Q_1, \ldots, Q_r$ as zero matrices, and $P_{\rm stab}$ as the solution of the Lyapunov equation $A_{\rm stab}^T P_{\rm stab} + P_{\rm stab} A_{\rm stab} + I = 0$.

In summary, the original LMI (2.23) has one symmetric matrix of size k as variable (i.e., the dimension of the original variable x is $m = k(k+1)/2$). The reduced LMI (2.25) has as variable the symmetric matrices $P_1, \ldots, P_{r+1}, P_{\rm stab}$ and the skew-symmetric matrices $Q_1, \ldots, Q_r$, so the dimension of the variable z in the reduced LMI is

$$p = \sum_{i=1}^{r} k_i^2 + \frac{k_{r+1}(k_{r+1}+1)}{2} + \frac{s(s+1)}{2} < m.$$

The original LMI (2.23) involves a matrix inequality of size $n = 2k$ (i.e., two inequalities of size k). The reduced LMI (2.25) involves a matrix inequality of size $k + s < n$.

2.6 Miscellaneous Results on Matrix Inequalities

2.6.1 Elimination of semidefinite terms

In the LMIP

$$\text{find } x \text{ such that } F(x) > 0, \tag{2.26}$$

we can eliminate any variables for which the corresponding coefficient in F is semidefinite. Suppose for example that $F_m \geq 0$ and has rank $r < n$. Roughly speaking, by taking x_m extremely large, we can ensure that F is positive-definite on the range of F_m, so the LMIP reduces to making F positive-definite on the nullspace of F_m.

More precisely, let $F_m = UU^T$ with U full-rank, and let $\tilde{U}$ be an orthogonal complement of U, e.g., $U^T\tilde{U} = 0$ and $[U\ \tilde{U}]$ is of maximum rank (which in this case just means that $[U\ \tilde{U}]$ is nonsingular). Of course, we have

$$F(x) > 0 \iff \left[\tilde{U}\ U\right]^T F(x) \left[\tilde{U}\ U\right] > 0.$$

Since

$$\left[\tilde{U}\ U\right]^T F(x) \left[\tilde{U}\ U\right] = \begin{bmatrix} \tilde{U}^T\tilde{F}(\tilde{x})\tilde{U} & \tilde{U}^T\tilde{F}(\tilde{x})U \\ U^T\tilde{F}(\tilde{x})\tilde{U} & U^T\tilde{F}(\tilde{x})U + x_m(U^TU)^2 \end{bmatrix},$$

where $\tilde{x} = [x_1 \cdots x_{m-1}]^T$ and $\tilde{F}(\tilde{x}) \stackrel{\Delta}{=} F_0 + x_1F_1 + \cdots + x_{m-1}F_{m-1}$, we see that $F(x) > 0$ if and only if $\tilde{U}^T\tilde{F}\tilde{U}(\tilde{x}) > 0$, and x_m is large enough that

$$U^T\tilde{F}(\tilde{x})U + x_m(U^TU)^2 > U^T\tilde{F}(\tilde{x})\tilde{U}\left(\tilde{U}^T\tilde{F}(\tilde{x})\tilde{U}\right)^{-1}\tilde{U}^T\tilde{F}(\tilde{x})U.$$

Therefore, the LMIP (2.26) is equivalent to

$$\text{find } x_1, \ldots, x_{m-1} \text{ such that } \tilde{U}^T\tilde{F}(x_1, \ldots, x_{m-1})\tilde{U} > 0.$$

2.6.2 Elimination of matrix variables

When a matrix inequality has some variables that appear in a certain form, we can derive an equivalent inequality without those variables. Consider

$$G(z) + U(z)XV(z)^T + V(z)X^TU(z)^T > 0, \tag{2.27}$$

where the vector z and the matrix X are (independent) variables, and $G(z) \in \mathbf{R}^{n\times n}$, $U(z)$ and $V(z)$ do not depend on X. Matrix inequalities of the form (2.27) arise in the controller synthesis problems described in Chapter 7.

Suppose that for every z, $\tilde{U}(z)$ and $\tilde{V}(z)$ are orthogonal complements of $U(z)$ and $V(z)$ respectively. Then (2.27) holds for some X and $z = z_0$ if and only if the inequalities

$$\tilde{U}(z)^TG(z)\tilde{U}(z) > 0, \qquad \tilde{V}(z)^TG(z)\tilde{V}(z) > 0 \tag{2.28}$$

hold with $z = z_0$. In other words, feasibility of the matrix inequality (2.27) with variables X and z is equivalent to the feasibility of (2.28) with variable z; we have eliminated the matrix variable X from (2.27) to form (2.28). Note that if $U(z)$ or $V(z)$ has rank n for all z, then the first or second inequality in (2.28) disappears. We prove this lemma in the Notes and References.

We can express (2.28) in another form using Finsler's lemma (see the Notes and References):

$$G(z) - \sigma U(z)U(z)^T > 0, \qquad G(z) - \sigma V(z)V(z)^T > 0$$

for some $\sigma \in \mathbf{R}$.

As an example, we will encounter in §7.2.1 the LMIP with LMI

$$Q > 0, \qquad AQ + QA^T + BY + Y^T B^T < 0, \tag{2.29}$$

where Q and Y are the variables. This LMIP is equivalent to the LMIP with LMI

$$Q > 0, \qquad AQ + QA^T < \sigma BB^T,$$

where the variables are Q and $\sigma \in \mathbf{R}$. It is also equivalent to the LMIP

$$Q > 0, \qquad \tilde{B}^T \left(AQ + QA^T\right) \tilde{B} < 0,$$

with variable Q, where $\tilde{B}$ is any matrix of maximum rank such that $\tilde{B}^T B = 0$. Thus we have eliminated the variable Y from (2.29) and reduced the size of the matrices in the LMI.

2.6.3 The $\mathcal{S}$-procedure

We will often encounter the constraint that some quadratic function (or quadratic form) be negative whenever some other quadratic functions (or quadratic forms) are all negative. In some cases, this constraint can be expressed as an LMI in the data defining the quadratic functions or forms; in other cases, we can form an LMI that is a conservative but often useful approximation of the constraint.

The $\mathcal{S}$-procedure for quadratic functions and nonstrict inequalities

Let $F_0, \ldots, F_p$ be quadratic functions of the variable $\zeta \in \mathbf{R}^n$:

$$F_i(\zeta) \stackrel{\Delta}{=} \zeta^T T_i \zeta + 2u_i^T \zeta + v_i, \quad i = 0, \ldots, p,$$

where $T_i = T_i^T$. We consider the following condition on $F_0, \ldots, F_p$:

$$F_0(\zeta) \geq 0 \text{ for all } \zeta \text{ such that } F_i(\zeta) \geq 0, \quad i = 1, \ldots, p. \tag{2.30}$$

Obviously if

$$\begin{array}{l} \text{there exist } \tau_1 \geq 0, \ldots, \tau_p \geq 0 \ \text{ such that} \\ \text{for all } \zeta, \quad F_0(\zeta) - \displaystyle\sum_{i=1}^{p} \tau_i F_i(\zeta) \geq 0, \end{array} \tag{2.31}$$

then (2.30) holds. It is a nontrivial fact that when $p = 1$, the converse holds, provided that there is some ζ_0 such that $F_1(\zeta_0) > 0$.

> **Remark**: If the functions F_i are affine, then (2.31) and (2.30) are equivalent; this is the Farkas lemma.

Note that (2.31) can be written as

$$\begin{bmatrix} T_0 & u_0 \\ u_0^T & v_0 \end{bmatrix} - \sum_{i=1}^{p} \tau_i \begin{bmatrix} T_i & u_i \\ u_i^T & v_i \end{bmatrix} \geq 0.$$

The $\mathcal{S}$-procedure for quadratic forms and strict inequalities

We will use another variation of the $\mathcal{S}$-procedure, which involves quadratic forms and strict inequalities. Let $T_0, \ldots, T_p \in \mathbf{R}^{n \times n}$ be symmetric matrices. We consider the following condition on $T_0, \ldots, T_p$:

$$\zeta^T T_0 \zeta > 0 \text{ for all } \zeta \neq 0 \text{ such that } \zeta^T T_i \zeta \geq 0, \quad i = 1, \ldots, p. \tag{2.32}$$

It is obvious that if

$$\text{there exists } \tau_1 \geq 0, \ldots, \tau_p \geq 0 \text{ such that } T_0 - \sum_{i=1}^{p} \tau_i T_i > 0, \tag{2.33}$$

then (2.32) holds. It is a nontrivial fact that when $p = 1$, the converse holds, provided that there is some ζ_0 such that $\zeta_0^T T_1 \zeta_0 > 0$. Note that (2.33) is an LMI in the variables T_0 and $\tau_1, \ldots, \tau_p$.

Remark: The first version of the $\mathcal{S}$-procedure deals with nonstrict inequalities and quadratic functions that may include constant and linear terms. The second version deals with strict inequalities and quadratic forms only, i.e., quadratic functions without constant or linear terms.

Remark: Suppose that T_0, u_0 and v_0 depend affinely on some parameter ν. Then the condition (2.30) is convex in ν. This does not, however, mean that the problem of checking whether there exists ν such that (2.30) holds has low complexity. On the other hand, checking whether (2.31) holds for some ν is an LMIP in the variables ν and $\tau_1, \ldots, \tau_p$. Therefore, this problem has low complexity. See the Notes and References for further discussion.

$\mathcal{S}$-procedure example

In Chapter 5 we will encounter the following constraint on the variable P:

$$\begin{array}{l} \text{for all } \xi \neq 0 \text{ and } \pi \text{ satisfying } \pi^T \pi \leq \xi^T C^T C \xi, \\ \begin{bmatrix} \xi \\ \pi \end{bmatrix}^T \begin{bmatrix} A^T P + PA & PB \\ B^T P & 0 \end{bmatrix} \begin{bmatrix} \xi \\ \pi \end{bmatrix} < 0. \end{array} \tag{2.34}$$

Applying the second version of the $\mathcal{S}$-procedure, (2.34) is *equivalent* to the existence of $\tau \geq 0$ such that

$$\begin{bmatrix} A^T P + PA + \tau C^T C & PB \\ B^T P & -\tau I \end{bmatrix} < 0.$$

Thus the problem of finding $P > 0$ such that (2.34) holds can be expressed as an LMIP (in P and the scalar variable τ).

2.7 Some LMI Problems with Analytic Solutions

There are analytic solutions to several LMI problems of special form, often with important system and control theoretic interpretations. We briefly describe some of these results in this section. At the same time we introduce and define several important terms from system and control theory.

2.7.1 Lyapunov's inequality

We have already mentioned the LMIP associated with Lyapunov's inequality, i.e.

$$P > 0, \qquad A^T P + PA < 0$$

where P is variable and $A \in \mathbf{R}^{n\times n}$ is given. Lyapunov showed that this LMI is feasible if and only the matrix A is stable, i.e., all trajectories of $\dot{x} = Ax$ converge to zero as $t \to \infty$, or equivalently, all eigenvalues of A must have negative real part. To solve this LMIP, we pick any $Q > 0$ and solve the Lyapunov equation $A^T P + PA = -Q$, which is nothing but a set of $n(n+1)/2$ linear equations for the $n(n+1)/2$ scalar variables in P. This set of linear equations will be solvable and result in $P > 0$ if and only if the LMI is feasible. In fact this procedure not only finds *a* solution when the LMI is feasible; it parametrizes *all* solutions as Q varies over the positive-definite cone.

2.7.2 The positive-real lemma

Another important example is given by the positive-real (PR) lemma, which yields a "frequency-domain" interpretation for a certain LMIP, and under some additional assumptions, a numerical solution procedure via Riccati equations as well. We give a simplified discussion here, and refer the reader to the References for more complete statements.

The LMI considered is:

$$P > 0, \qquad \begin{bmatrix} A^T P + PA & PB - C^T \\ B^T P - C & -D^T - D \end{bmatrix} \leq 0, \tag{2.35}$$

where $A \in \mathbf{R}^{n\times n}$, $B \in \mathbf{R}^{n\times p}$, $C \in \mathbf{R}^{p\times n}$, and $D \in \mathbf{R}^{p\times p}$ are given, and the matrix $P = P^T \in \mathbf{R}^{n\times n}$ is the variable. (We will encounter a variation on this LMI in §6.3.3.) Note that if $D + D^T > 0$, the LMI (2.35) is equivalent to the quadratic matrix inequality

$$A^T P + PA + (PB - C^T)(D + D^T)^{-1}(PB - C^T)^T \leq 0. \tag{2.36}$$

We assume for simplicity that A is stable and the system (A, B, C) is minimal.

The link with system and control theory is given by the following result. The LMI (2.35) is feasible if and only if the linear system

$$\dot{x} = Ax + Bu, \qquad y = Cx + Du \tag{2.37}$$

is *passive*, i.e.,

$$\int_0^T u(t)^T y(t)\, dt \geq 0$$

for all solutions of (2.37) with $x(0) = 0$.

Passivity can also be expressed in terms of the *transfer matrix* of the linear system (2.37), defined as

$$H(s) \stackrel{\Delta}{=} C(sI - A)^{-1}B + D$$

for $s \in \mathbf{C}$. Passivity is equivalent to the transfer matrix H being *positive-real*, which means that

$$H(s) + H(s)^* \geq 0 \qquad \text{for all} \quad \mathbf{Re}\, s > 0.$$

When $p = 1$ this condition can be checked by various graphical means, e.g., plotting the curve given by the real and imaginary parts of $H(i\omega)$ for $\omega \in \mathbf{R}$ (called the *Nyquist plot* of the linear system). Thus we have a graphical, frequency-domain condition for feasibility of the LMI (2.35). This approach was used in much of the work in the 1960s and 1970s described in §1.2.

With a few further technical assumptions, including $D + D^T > 0$, the LMI (2.35) can be solved by a method based on Riccati equations and Hamiltonian matrices. With these assumptions the LMI (2.35) is feasible if and only if there exists a real matrix $P = P^T$ satisfying the ARE

$$A^T P + PA + (PB - C^T)(D + D^T)^{-1}(PB - C^T)^T = 0, \tag{2.38}$$

which is just the quadratic matrix inequality (2.36) with equality substituted for inequality. Note that $P > 0$.

To solve the ARE (2.38) we first form the associated Hamiltonian matrix

$$M = \begin{bmatrix} A - B(D + D^T)^{-1}C & B(D + D^T)^{-1}B^T \\ -C^T(D + D^T)^{-1}C & -A^T + C^T(D + D^T)^{-1}B^T \end{bmatrix}.$$

Then the system (2.37) is passive, or equivalently, the LMI (2.35) is feasible, if and only if M has no pure imaginary eigenvalues. This fact can be used to form a Routh–Hurwitz (Sturm) type test for passivity as a set of polynomial inequalities in the data A, B, C, and D.

When M has no pure imaginary eigenvalues we can construct a solution $P_{\rm are}$ as follows. Pick $V \in \mathbf{R}^{2n \times n}$ so that its range is a basis for the stable eigenspace of M, e.g., $V = [v_1 \cdots v_n]$ where $v_1, \ldots, v_n$ are a set of independent eigenvectors of M associated with its n eigenvalues with negative real part. Partition V as

$$V = \begin{bmatrix} V_1 \\ V_2 \end{bmatrix},$$

where V_1 and V_2 are square; then set $P_{\rm are} = V_2 V_1^{-1}$. The solution $P_{\rm are}$ thus obtained is the *minimal* element among the set of solutions of (2.38): if $P = P^T$ satisfies (2.38), then $P \geq P_{\rm are}$. Much more discussion of this method, including the precise technical conditions, can be found in the References.

2.7.3 The bounded-real lemma

The same results appear in another important form, the *bounded-real lemma.* Here we consider the LMI

$$P > 0, \qquad \begin{bmatrix} A^T P + PA + C^T C & PB + C^T D \\ B^T P + D^T C & D^T D - I \end{bmatrix} \leq 0. \tag{2.39}$$

where $A \in \mathbf{R}^{n \times n}$, $B \in \mathbf{R}^{n \times p}$, $C \in \mathbf{R}^{p \times n}$, and $D \in \mathbf{R}^{p \times p}$ are given, and the matrix $P = P^T \in \mathbf{R}^{n \times n}$ is the variable. For simplicity we assume that A is stable and (A, B, C) is minimal.

This LMI is feasible if and only the linear system (2.37) is *nonexpansive,* i.e.,

$$\int_0^T y(t)^T y(t)\, dt \leq \int_0^T u(t)^T u(t)\, dt$$

for all solutions of (2.37) with $x(0) = 0$, This condition can also be expressed in terms of the transfer matrix H. Nonexpansivity is equivalent to the transfer matrix H satisfying the *bounded-real* condition, i.e.,

$$H(s)^*H(s) \leq I \qquad \text{for all} \quad \mathbf{Re}\, s > 0.$$

This is sometimes expressed as $\|H\|_\infty \leq 1$ where

$$\|H\|_\infty \stackrel{\Delta}{=} \sup\{\ \|H(s)\|\ \mid\ \mathbf{Re}\, s > 0\}$$

is called the $\mathbf{H}_\infty$ *norm* of the transfer matrix H.

This condition is easily checked graphically, e.g., by plotting $\|H(i\omega)\|$ versus $\omega \in \mathbf{R}$ (called a *singular value plot* for $p > 1$ and a *Bode magnitude plot* for $p = 1$).

Once again we can relate the LMI (2.39) to an ARE. With some appropriate technical conditions (see the Notes and References) including $D^TD < I$, the LMI (2.39) is feasible if and only the ARE

$$A^TP + PA + C^TC + (PB + C^TD)(I - D^TD)^{-1}(PB + C^TD)^T = 0 \qquad (2.40)$$

has a real solution $P = P^T$. Once again this is the quadratic matrix inequality associated with the LMI (2.39), with equality instead of the inequality.

We can solve this equation by forming the Hamiltonian matrix

$$M = \begin{bmatrix} A + B(I - D^TD)^{-1}D^TC & B(I - D^TD)^{-1}B^T \\ -C^T(I - DD^T)^{-1}C & -A^T - C^TD(I - D^TD)^{-1}B^T \end{bmatrix}.$$

Then the system (2.37) is nonexpansive, or equivalently, the LMI (2.39) is feasible, if and only if M has no imaginary eigenvalues. In this case we can find the minimal solution to (2.40) as described in §2.7.2.

2.7.4 Others

Several other LMI problems have analytic solutions, e.g., the ones that arise in the synthesis of state-feedback for linear systems, or estimator gains for observers. As a simple example consider the LMI

$$P > 0, \qquad AP + PA^T < BB^T$$

where $A \in \mathbf{R}^{n\times n}$ and $B \in \mathbf{R}^{n\times p}$ are given and P is the variable. This LMI is feasible if and only if the pair (A, B) is stabilizable, i.e., there exists a matrix $K \in \mathbf{R}^{p\times n}$ such that $A + BK$ is stable. There are simple methods for determining whether this is the case, and when it is, of constructing an appropriate P. We will describe this and other results in Chapter 7.

Several standard results from linear algebra can also be interpreted as analytic solutions to certain LMI problems of special form. For example, it is well-known that among all symmetric positive-definite matrices with given diagonal elements, the diagonal matrix has maximum determinant. This gives an analytic solution to a special (and fairly trivial) CP (see §3.5).

Notes and References

LMIs

The term "Linear Matrix Inequality" was coined by J. C. Willems and is widely used now. As we mentioned in §1.2, the term was used in several papers from the 1970s to refer to the specific LMI (1.3). The term is also consistent with the title of an early paper by Bellman and Ky Fan: *On Systems of Linear Inequalities in Hermitian Matrix Variables* [BF63] (see below).

A more accurate term might be "Affine Matrix Inequality", since the matrix is an affine and not linear function of the variable. We think of the term "Linear Matrix Inequality" as analogous to the term "Linear Inequalities" used to describe $a_i^T x \leq b_i$, $i = 1, \ldots, p$.

When $m = 1$, an LMI is nothing but a matrix pencil. The theory of matrix pencils is very old and quite complete, and there is an extensive literature (see, e.g., Golub and Van Loan [GL89] and the references therein). For the case of $m = 1$ all of our standard problems are readily solved, e.g., by simultaneous diagonalization of F_0 and F_1 by an appropriate congruence. The case $m = 1$ does not arise on its own in any interesting problem from system or control theory, but it does arise in the "line-search (sub)problem", i.e., when we restrict one of our standard problems to a specific line. In this context the variable represents the "step length" to be taken in an iteration of an algorithm, and it useful to know that such problems can be solved extremely efficiently.

Schur complements for nonstrict inequalities

The Schur complement result of section §2.1 can be generalized to nonstrict inequalities as follows. Suppose Q and R are symmetric. The condition

$$\begin{bmatrix} Q & S \\ S^T & R \end{bmatrix} \geq 0 \tag{2.41}$$

is equivalent to

$$R \geq 0, \qquad Q - SR^\dagger S^T \geq 0, \qquad S(I - RR^\dagger) = 0,$$

where $R^\dagger$ denotes the Moore–Penrose inverse of R.

To see this, let U be an orthogonal matrix that diagonalizes R, so that

$$U^T R U = \begin{bmatrix} \Sigma & 0 \\ 0 & 0 \end{bmatrix},$$

where $\Sigma > 0$ and diagonal. Inequality (2.41) holds if and only if

$$\begin{bmatrix} I & 0 \\ 0 & U^T \end{bmatrix} \begin{bmatrix} Q & S \\ S^T & R \end{bmatrix} \begin{bmatrix} I & 0 \\ 0 & U \end{bmatrix} = \begin{bmatrix} Q & S_1 & S_2 \\ S_1^T & \Sigma & 0 \\ S_2^T & 0 & 0 \end{bmatrix} \geq 0,$$

where $[S_1 \ \ S_2] = SU$, with appropriate partitioning. We must then have $S_2 = 0$, which holds if and only if $S(I - RR^\dagger) = 0$, and

$$\begin{bmatrix} Q & S_1 \\ S_1^T & \Sigma \end{bmatrix} \geq 0,$$

which holds if and only if $Q - SR^\dagger S^T \geq 0$.

Formulating convex problems in terms of LMIs

The idea that LMIs can be used to represent a wide variety of convex constraints can be found in Nesterov and Nemirovskii's report [NN90B] (which is really a preliminary version of their book [NN94]) and software manual [NN90A]. They formalize the idea of a "positive-definite representable" function; see §5.3, §5.4, and §6.4 of their book [NN94]. The idea is also discussed in the article [ALI92B] and thesis [ALI91] of Alizadeh. In [BE93], Boyd and El Ghaoui give a list of quasiconvex functions that can be represented as generalized eigenvalues of matrices that depend affinely on the variable.

Infeasibility criterion for LMIs

The LMI $F(x) > 0$ is infeasible means the affine set $\{F(x) \mid x \in \mathbf{R}^m\}$ does not intersect the positive-definite cone. From convex analysis, this is equivalent to the existence of a linear functional ψ that is positive on the positive-definite cone and nonpositive on the affine set of matrices. The linear functionals that are positive on the positive-definite cone are of the form $\psi(F) = \mathbf{Tr}\, GF$, where $G \geq 0$ and $G \neq 0$. From the fact that ψ is nonpositive on the affine set $\{F(x)|x \in \mathbf{R}^m\}$, we can conclude that $\mathbf{Tr}\, GF_i = 0$, $i = 1, \ldots, m$ and $\mathbf{Tr}\, GF_0 \leq 0$. These are precisely the conditions for infeasibility of an LMI that we mentioned in §2.2.1.

For the special case of multiple Lyapunov inequalities, these conditions are given Bellman and Ky Fan [BF63] and Kamenetskii and Pyatnitskii [KP87A, KP87B].

It is straightforward to derive optimality criteria for the other problems, using convex analysis. Some general references for convex analysis are the books [ROC70, ROC82] and survey article [ROC93] by Rockafellar. The recent text [HUL93] gives a good overview of convex analysis; LMIs are used as examples in several places.

Complexity of convex optimization

The important role of convexity in optimization is fairly widely known, but perhaps not well enough appreciated, at least outside the former Soviet Union. In a standard (Western) treatment of optimization, our standard problems LMIP, EVP, GEVP, and CP would be considered very difficult since they are nondifferentiable and nonlinear. Their convexity properties, however, make them tractable, both in theory and in practice.

In [ROC93, P194], Rockafellar makes the important point:

> One distinguishing idea which dominates many issues in optimization theory is convexity ... An important reason is the fact that when a convex function is minimized over a convex set every locally optimal solution is global. Also, first-order necessary conditions turn out to be sufficient. A variety of other properties conducive to computation and interpretation of solutions ride on convexity as well. In fact the great watershed in optimization isn't between linearity and nonlinearity, but convexity and nonconvexity.

Detailed discussion of the (low) complexity of convex optimization problems can be found in the books by Grötschel, Lovász, Schrijver [GLS88], Nemirovskii and Yudin [NY83], and Vavasis [VAV91].

Complete and detailed worst-case complexity analyses of several algorithms for our standard problems can be found in Chapters 3, 4, and 6 of Nesterov and Nemirovskii [NN94].

Ellipsoid algorithm

The ellipsoid algorithm was developed by Shor, Nemirovskii, and Yudin in the 1970s [SHO85, NY83]. It was used by Khachiyan in 1979 to prove that linear programs can be solved in polynomial-time [KHA79, GL81, GLS88]. Discussion of the ellipsoid method, as well as several extensions and variations, can be found in [BGT81] and [BB91, CH14]. A detailed history of its development, including English and Russian references, appears in chapter 3 of Akgül [AKG84].

Optimization problems involving LMIs

One of the earliest papers on LMIs is also, in our opinion, one of the best: *On systems of linear inequalities in Hermitian matrix variables*, by Bellman and Ky Fan [BF63]. In this paper we find many results, e.g., a duality theory for the multiple Lyapunov inequality EVP and a theorem of the alternative for the multiple Lyapunov inequality LMIP. The paper concentrates on the associated mathematics: there is no discussion of how to solve such problems, or any potential applications, although they do comment that such inequality systems arise in Lyapunov theory. Given Bellman's legendary knowledge of system and control theory, and especially Lyapunov theory, it is tempting to conjecture that Bellman was aware of the potential applications. But so far we have found no evidence for this.

Relevant work includes Cullum et al. [CDW75], Craven and Mond [CM81], Polak and Wardi [PW82], Fletcher [FLE85], Shapiro [SHA85], Friedland et al. [FNO87], Goh and Teo [GT88], Panier [PAN89], Allwright [ALL89], Overton and Womersley [OVE88, OVE92, OW93, OW92], Ringertz [RIN91], Fan and Nekooie [FN92, FAN93] and Hiriart–Urruty and Ye [HUY92].

LMIPs are solved using various algorithms for convex optimization in Boyd and Yang [BY89], Pyatnitskii and Skorodinskii [PS83], Kamenetskii and Pyatnitskii [KP87A, KP87B].

A survey of methods for solving problems involving LMIs used by researchers in control theory can be found in the paper by Beck [BEC91]. The software packages [BDG+91] and [CS92A] use convex programming to solve many robust control problems. Boyd [BOY94] outlines how interior-point convex optimization algorithms can be used to build robust control software tools. Convex optimization problems involving LMIs have been used in control theory since about 1985: Gilbert's method [GIL66] was used by Doyle [DOY82] to solve a diagonal scaling problem; another example is the MUSOL program of Fan and Tits [FT86].

Among other methods used in control to solve problems involving multiple LMIs is the "method of alternating convex projections" described in the article by Grigoriadis and Skelton [GS92]. This method is essentially a relaxation method, and requires an analytic expression for the projection onto the feasible set of each LMI. It is not a polynomial-time algorithm; its complexity depends on the problem data, unlike the ellipsoid method or the interior-point methods described in [NN94], whose complexities depend on the problem size *only*. However, the alternating projections method is reported to work well in many cases.

Interior-point methods for LMI problems

Interior-point methods for various LMI problems have recently been developed by several researchers. The first were Nesterov and Nemirovskii [NN88, NN90B, NN90A, NN91A, NN94, NN93]; others include Alizadeh [ALI92B, ALI91, ALI92A], Jarre [JAR93C], Vandenberghe and Boyd [VB93B], Rendl, Vanderbei, and Wolkowocz [RVW93], and Yoshise [YOS94].

Of course, general interior-point methods (and the method of centers in particular) have a long history. Early work includes the book by Fiacco and McCormick [FM68], the method of centers described by Huard et al. [LH66, HUA67], and Dikin's interior-point method for linear programming [DIK67]. Interest in interior-point methods surged in 1984 when Karmarkar [KAR84] gave his interior-point method for solving linear programs, which appears to have very good practical performance as well as a good worst-case complexity bound. Since the publication of Karmarkar's paper, many researchers have studied interior-point methods for linear and quadratic programming. These methods are often described in such a way that extensions to more general (convex) constraints and objectives are not clear. However, Nesterov and Nemirovskii have developed a theory of interior-point methods that applies to more general convex programming problems, and in particular, *every* problem that arises in this book; see [NN94]. In particular, they derive complexity bounds for many different interior-point algorithms, including the method of centers, Nemirovskii's projective algorithm and primal-dual methods. The most efficient algorithms seem to be Nemirovskii's projective algorithm and primal-dual methods (for the case of linear programs, see [MEH91]).

Other recent articles that consider interior-point methods for more general convex programming include Sonnevend [SON88], Jarre [JAR91, JAR93B], Kortanek et al. [KPY91], Den Hertog, Roos, and Terlaky [DRT92], and the survey by Wright [WRI92].

Interior-point methods for GEVPs are described in Boyd and El Ghaoui [BE93] (variation on the method of centers), and Nesterov and Nemirovskii [NN91B] and [NN94, §4.4] (a variation on Nemirovskii's projective algorithm). Since GEVPs are not convex problems, devising a reliable stopping criterion is more challenging than for the convex problems LMIP, EVP, and CP. A detailed complexity analysis (in particular, a statement and proof of the polynomial-time complexity of GEVPs) is given in [NN91B, NN94]. See also [JAR93A].

Several researchers have recently studied the possibility of switching from an interior-point method to a quadratically convergent local method in order to improve on the final convergence; see [OVE92, OW93, OW92, FN92, NF92].

Interior-point methods for CPs and other related extremal ellipsoid volume problems can be found in [NN94, §6.5].

Interior-point methods and problem structure

Many researchers have developed algorithms that take advantage of the special structure of the least-squares problems arising in interior-point methods for linear programming. As far as we know, the first (and so far, only) interior-point algorithm that takes advantage of the special (Lyapunov) structure of an LMI problem arising in control is described in Vandenberghe and Boyd [VB93B, VB93A]. Nemirovskii's projective method can also take advantage of such structure; these two algorithms appear to be the most efficient algorithms developed so far for solving the LMIs that arise in control theory.

Software for solving LMI problems

Gahinet and Nemirovskii have recently developed a software package called LMI-Lab [GN93] based on an earlier FORTRAN code [NN90A], which allows the user to describe an LMI problem in a high-level symbolic form (not unlike the formulas that appear throughout this book!). LMI-Lab then solves the problem using Nemirovskii's projective algorithm, taking advantage of some of the problem structure (e.g., block structure, diagonal structure of some of the matrix variables).

Recently, El Ghaoui has developed another software package for solving LMI problems. This noncommercial package, called LMI-tool, can be used with MATLAB. It is available via anonymous ftp (for more information, send mail to `elghaoui@ensta.fr`). Another version of LMI-tool, developed by Nikoukhah and Delebecque, is available for use with the MATLAB-like freeware package SCILAB; in this version, LMI-tool has been interfaced with Nemirovskii's projective code. SCILAB can be obtained via anonymous ftp (for more information, send mail to `Scilab@inria.fr`).

A commercial software package that solves a few specialized control system analysis and design problems via LMI formulation, called OPTIN, was recently developed by Olas and Associates; see [OS93, OLA94].

Reduction to a strictly feasible LMI

In this section we prove the following statement. Let $F_0, \ldots, F_m \in \mathbf{R}^{n\times n}$ be symmetric matrices. Then there is a matrix $A \in \mathbf{R}^{m\times p}$ with $p \le m$, a vector $b \in \mathbf{R}^m$, and symmetric matrices $\tilde{F}_0, \ldots, \tilde{F}_p \in \mathbf{R}^{q\times q}$ with $q \le n$ such that:

$$F(x) \ge 0 \text{ if and only if } x = Az + b \text{ for some } z \in \mathbf{R}^p \text{ and } \tilde{F}(z) \triangleq \tilde{F}_0 + \sum_{i=1}^{p} z_i \tilde{F}_i \ge 0.$$

In addition, if the LMI $F(x) \ge 0$ is feasible, then the LMI $\tilde{F}(z) \ge 0$ is strictly feasible.

Consider the LMI

$$F(x) = F_0 + \sum_{i=1}^{m} x_i F_i \geq 0, \tag{2.42}$$

where $F_i \in \mathbf{R}^{n\times n}$, $i = 0, \ldots, m$. Let $\mathbf{X}$ denote the feasible set $\{x \mid F(x) \geq 0\}$.

If $\mathbf{X}$ is empty, there is nothing to prove; we can take $p = m$, $A = I$, $b = 0$, $\tilde{F}_0 = F_0, \ldots, \tilde{F}_p = F_p$.

Henceforth we assume that $\mathbf{X}$ is nonempty. If $\mathbf{X}$ is a singleton, say $\mathbf{X} = \{x_0\}$, then with $p = 1$, $A = 0$, $b = x_0$, $\tilde{F}_0 = 1$, and $\tilde{F}_1 = 0$, the statement follows.

Now consider the case when $\mathbf{X}$ is neither empty nor a singleton. Then, there is an affine subspace $\mathcal{A}$ of minimal dimension $p \geq 1$ that contains $\mathbf{X}$. Let $a_1, \ldots a_p$ a basis for the linear part of $\mathcal{A}$. Then every $x \in \mathcal{A}$ can be written as $x = Az + b$ where $A = [a_1 \ \cdots \ a_p]$ is full-rank and $b \in \mathbf{R}^m$. Defining $G_0 = F(b)$, $G_i = F(a_i) - F_0$, $i = 1, \ldots, p$, and $G(z) = G_0 + \sum_{i=1}^{p} z_i G_i$, we see that $F(x) \geq 0$ if and only if there exists $z \in \mathbf{R}^p$ satisfying $x = Az + b$ and $G(z) \geq 0$.

Let $\mathbf{Z} \stackrel{\Delta}{=} \{z \in \mathbf{R}^p \mid G(z) \geq 0\}$. By construction, $\mathbf{Z}$ has nonempty interior. Let z_0 a point in the interior of $\mathbf{Z}$ and let v lie in the nullspace of $G(z_0)$. Now, $v^T G(z) v$ is a nonnegative affine function of $z \in \mathbf{Z}$ and is zero at an interior point of $\mathbf{Z}$. Therefore, it is identically zero over $\mathbf{Z}$, and hence over $\mathbf{R}^p$. Therefore, v belongs to the intersection $\mathcal{B}$ of the nullspaces of the G_i, $i = 0, \ldots, p$. Conversely, any v belonging to $\mathcal{B}$ will satisfy $v^T G(z) v = 0$ for any $z \in \mathbf{R}^p$. $\mathcal{B}$ may therefore be interpreted as the "constant" nullspace of the LMI (2.42). Let q be the dimension of $\mathcal{B}$.

If $q = n$ (i.e., $G_0 = \cdots = G_p = 0$), then obviously $F(x) \geq 0$ if and only if $x = Az + b$. In this case, the statement is satisfied with $\tilde{F}_0 = I \in \mathbf{R}^{p\times p}$, $\tilde{F}_i = 0$, $i = 1, \ldots, p$.

If $q < n$, let $v_1, \ldots, v_q$ a basis for $\mathcal{B}$. Complete the basis $v_1, \ldots, v_q$ by $v_{q+1}, \ldots, v_n$ to obtain a basis of $\mathbf{R}^n$. Define $U = [v_{q+1} \ \cdots \ v_n]$, $\tilde{F}_i = U^T G_i U$, $i = 1, \ldots, p$ and $\tilde{F}(z) = \tilde{F}_0 + \sum_{i=1}^{p} z_i \tilde{F}_i$. Then, the LMI $G(z) \geq 0$ is equivalent to the LMI $\tilde{F}(z) \geq 0$, and by construction, there exists z_0 such that $\tilde{F}(z_0) > 0$. This concludes the proof.

Elimination procedure for matrix variables

We now prove the matrix elimination result stated in §2.6.2: Given G, U, V, there exists X such that

$$G + UXV^T + VX^TU^T > 0 \tag{2.43}$$

if and only if

$$\tilde{U}^T G \tilde{U} > 0, \qquad \tilde{V}^T G \tilde{V} > 0 \tag{2.44}$$

holds, where $\tilde{U}$ and $\tilde{V}$ are orthogonal complements of U and V respectively.

It is obvious that if (2.43) holds for some X, so do inequalities (2.44). Let us now prove the converse.

Suppose that inequalities (2.44) are feasible. Suppose now that inequality (2.43) is not feasible for any X. In other words, suppose that

$$G + UXV^T + VX^TU^T \not> 0,$$

for every X. By duality, this is equivalent to the condition

$$\text{there exists } Z \neq 0 \text{ with } Z \geq 0,\ V^T Z U = 0, \text{ and } \mathbf{Tr}\, GZ \leq 0. \tag{2.45}$$

Now let us show that $Z \geq 0$ and $V^T ZU = 0$ imply that

$$Z = \tilde{V}HH^T\tilde{V}^T + \tilde{U}KK^T\tilde{U}^T \tag{2.46}$$

for some matrices H, K (at least one of which is nonzero, since otherwise $Z = 0$). This will finish our proof, since (2.44) implies that $\mathbf{Tr}\, GZ > 0$ for Z of the form (2.46), which contradicts $\mathbf{Tr}\, GZ \leq 0$ in (2.45).

Let R be a Cholesky factor of Z, i.e. $Z = RR^T$. The condition $V^T ZU = 0$ is equivalent to

$$V^T RR^T U = (V^T R)(U^T R)^T = 0.$$

This means that there exists a unitary matrix T, and matrices M and N such that

$$V^T RT = [0 \;\; M] \qquad \text{and} \qquad U^T RT = [N \;\; 0],$$

where the number of columns of M and N add up to that of R. In other words, RT can be written as

$$RT = [A \;\; B]$$

for some matrices A, B such that $V^T A = 0$, $U^T B = 0$. From the definition of $\tilde{V}$ and $\tilde{U}$, matrices A, B can be written $A = \tilde{V}H$, $B = \tilde{U}K$ for some matrices H, K, and we have

$$Z = RR^T = (RT)(RT)^T = AA^T + BB^T = \tilde{V}HH^T\tilde{V}^T + \tilde{U}KK^T\tilde{U}^T,$$

which is the desired result.

The equivalence between the conditions

$$\tilde{U}(z)^T G(z)\tilde{U}(z) > 0$$

and

$$G(z) - \sigma U(z)U(z)^T > 0 \text{ for some real scalar } \sigma$$

is through Finsler's lemma [FIN37] (see also §3.2.6 of [SCH73]), which states that if $x^T Qx > 0$ for all nonzero x such that $x^T Ax = 0$, where Q and A are symmetric, real matrices, then there exists a real scalar such that $Q - \sigma A > 0$.

Finsler's lemma has also been directly used to eliminate variables in certain matrix inequalities (see for example [PH86, KR88, BPG89B]); it is closely related to the $\mathcal{S}$-procedure. We also note another result on elimination of matrix variables, due to Parrott [PAR78].

The elimination lemma is related to a matrix dilation problem considered in [DKW82], which was used in control problems in [PZPB91, PZP+92, GAH92, PAC94, IS93A, IWA93] (see also the Notes and References of Chapter 7).

The $\mathcal{S}$-procedure

The problem of determining if a quadratic form is nonnegative when other quadratic forms are nonnegative has been studied by mathematicians for at least seventy years. For a complete discussion and references, we refer the reader to the survey article by Uhlig [UHL79]. See also the book by Hestenes [HES81, P354-360] and Horn and Johnson [HJ91, P78-86] for proofs of various $\mathcal{S}$-procedure results.

The application of the $\mathcal{S}$-procedure to control problems dates back to 1944, when Lur'e and Postnikov used it to prove the stability of some particular nonlinear systems. The name "$\mathcal{S}$-procedure" was introduced much later by Aizerman and Gantmacher in their 1964 book [AG64]; since then Yakubovich has given more general formulations and corrected errors in some earlier proofs [YAK77, FY79].

The proof of the two $\mathcal{S}$-procedure results described in this chapter can be found in the articles by Yakubovich [FY79, YAK77, YAK73]. Recent and important developments about the $\mathcal{S}$-procedure can be found in [YAK92] and references therein.

Complexity of $\mathcal{S}$-procedure condition

Suppose that T_0, u_0, and v_0 depend affinely on some variable ν; the condition (2.30) is convex in ν. Indeed, for fixed ζ, the constraint $F_0(\zeta) \geq 0$ is a linear constraint on ν and the constraint (2.30) is simply an infinite number of these linear constraints. We might therefore imagine that the problem of checking whether there exists ν such that (2.30) holds has low complexity since it can be cast as a convex feasibility problem. This is wrong. In fact, merely verifying that the condition (2.30) holds for fixed data T_i, u_i, and v_i, is as hard as solving a general indefinite quadratic program, which is NP-complete. We could determine whether there exists ν such that (2.30) holds with polynomially many steps of the ellipsoid algorithm; the problem is that finding a cutting plane (which is required for each step of the ellipsoid algorithm) is itself an NP-complete problem.

Positive-real and bounded-real lemma

In 1961, Popov gave the famous Popov frequency-domain stability criterion for the absolute stability problem [POP62]. Popov's criterion could be checked via graphical means, by verifying that the Nyquist plot of the "linear part" of the nonlinear system was confined to a specific region in the complex plane. Yakubovich [YAK62, YAK64] and Kalman [KAL63A, KAL63B] established the connection between the Popov criterion and the existence of a positive-definite matrix satisfying certain matrix inequalities. The PR lemma is also known by various names such as the Yakubovich–Kalman–Popov–Anderson Lemma or the Kalman–Yakubovich Lemma. The PR lemma is now standard material, described in several books on control and systems theory, e.g., Narendra and Taylor [NT73], Vidyasagar [VID92, PP474–478], Faurre and Depeyrot [FD77], Anderson and Vongpanitlerd [AV73, CH5–7], Brockett [BRO70], and Willems [WIL70].

In its original form, the PR Lemma states that the transfer function $c(sI - A)^{-1}b$ of the single-input single-output minimal system (A, b, c) is positive-real, i.e.

$$\mathbf{Re}\, c(sI - A)^{-1}b \geq 0 \qquad \text{for all } \mathbf{Re}\, s > 0 \tag{2.47}$$

if and only if there exists $P > 0$ such that $A^TP + PA \leq 0$ and $Pb = c^T$. The condition (2.47) can be checked graphically.

Anderson [AND67, AND66B, AND73] extended the PR lemma to multi-input multi-output systems, and derived similar results for nonexpansive systems [AND66A]. Willems [WIL71B, WIL74A] described connections between the PR lemma, certain quadratic optimal control problems and the existence of symmetric solutions to the ARE; it is in [WIL71B] that we find Willems' quote on the role of LMIs in quadratic optimal control (see page 3).

The connections between passivity, LMIs and AREs can be summarized as follows. We consider A, B, C, and D such that all eigenvalues of A have negative real part, (A, B) is controllable and $D + D^T > 0$. The following statements are equivalent:

1. The system

$$\dot{x} = Ax + Bu, \qquad y = Cx + Du, \qquad x(0) = 0$$

is passive, i.e., satisfies

$$\int_0^T u(t)^T y(t)\,dt \geq 0$$

for all u and $T \geq 0$.

2. The transfer matrix $H(s) = C(sI - A)^{-1}B + D$ is positive-real, i.e.,
$$H(s) + H(s)^* \geq 0$$
for all s with $\mathbf{Re}\, s \geq 0$.
3. The LMI
$$\begin{bmatrix} A^TP + PA & PB - C^T \\ B^TP - C & -(D + D^T) \end{bmatrix} \leq 0 \tag{2.48}$$
in the variable $P = P^T$ is feasible.
4. There exists $P = P^T$ satisfying the ARE
$$A^TP + PA + (PB - C^T)(D + D^T)^{-1}(PB - C^T)^T = 0. \tag{2.49}$$
5. The sizes of the Jordan blocks corresponding to the pure imaginary eigenvalues of the Hamiltonian matrix
$$M = \begin{bmatrix} A - B(D + D^T)^{-1}C & B(D + D^T)^{-1}B^T \\ -C^T(D + D^T)^{-1}C & -A^T + C^T(D + D^T)^{-1}B^T \end{bmatrix}$$
are all even.

The equivalence of 1, 2, 3, and 4 can be found in Theorem 4 of [WIL71B] and also [WIL74A]. The equivalence of 5 is found in, for example, Theorem 1 of [LR80]. Origins of this result can be traced back to Reid [REI46] and Levin [LEV59]; it is explicitly stated in Potter [POT66]. An excellent discussion of this result and its connections with spectral factorization theory can be found in [AV73]; see also [FRA87, BBK89, ROB89]. Connections between the extremal points of the set of solutions of the LMI (2.48) and the solutions of the ARE (2.49) are explored in [BAD82]. The method for constructing $P_{\rm are}$ from the stable eigenspace of M, described in §2.7.2, is in [BBK89]; this method is an obvious variation of Laub's algorithm [LAU79, AL84], or Van Dooren's [DOO81]. A less stable numerical method can be found in [REI46, LEV59, POT66, AV73]. For other discussions of strict passivity, see the articles by Wen [WEN88], Lozano-Leal and Joshi [LLJ90]. It is also possible to check passivity (or nonexpansivity) of a system symbolically using Sturm methods, which yields Routh–Hurwitz like algorithms; see Siljak [SIL71, SIL73] and Boyd, Balakrishnan, and Kabamba [BBK89].

The PR lemma is used in many areas, e.g., interconnected systems [MH78] and stochastic processes and stochastic realization theory [FAU67, FAU73, AM74].

Chapter 3

Some Matrix Problems

3.1 Minimizing Condition Number by Scaling

The *condition number* of a matrix $M \in \mathbf{R}^{p\times q}$, with $p \geq q$, is the ratio of its largest and smallest singular values, i.e.,

$$\kappa(M) \triangleq \left(\frac{\lambda_{\max}(M^TM)}{\lambda_{\min}(M^TM)}\right)^{1/2}$$

for M full-rank, and $\kappa(M) = \infty$ otherwise. For square invertible matrices this reduces to $\kappa(M) = \|M\|\|M^{-1}\|$.

We consider the problem:

$$\begin{array}{ll} \text{minimize} & \kappa\,(LMR) \\ \text{subject to} & L \in \mathbf{R}^{p\times p}, \text{ diagonal and nonsingular} \\ & R \in \mathbf{R}^{q\times q}, \text{ diagonal and nonsingular} \end{array} \tag{3.1}$$

where L and R are the optimization variables, and the matrix $M \in \mathbf{R}^{p\times q}$ is given. We will show that this problem can be transformed into a GEVP. We assume without loss of generality that $p \geq q$ and M is full-rank.

Let us fix $\gamma > 1$. There exist nonsingular, diagonal $L \in \mathbf{R}^{p\times p}$ and $R \in \mathbf{R}^{q\times q}$ such that $\kappa(LMR) \leq \gamma$ if and only if there are nonsingular, diagonal $L \in \mathbf{R}^{p\times p}$ and $R \in \mathbf{R}^{q\times q}$ and $\mu > 0$ such that

$$\mu I \leq (LMR)^T(LMR) \leq \mu\gamma^2 I.$$

Since we can absorb the factor $1/\sqrt{\mu}$ into L, this is equivalent to the existence of nonsingular, diagonal $L \in \mathbf{R}^{p\times p}$ and $R \in \mathbf{R}^{q\times q}$ such that

$$I \leq (LMR)^T(LMR) \leq \gamma^2 I,$$

which is the same as

$$(RR^T)^{-1} \leq M^T(L^TL)M \leq \gamma^2(RR^T)^{-1}. \tag{3.2}$$

This is equivalent to the existence of diagonal $P \in \mathbf{R}^{p\times p}$, $Q \in \mathbf{R}^{q\times q}$ with $P > 0$, $Q > 0$ and

$$Q \leq M^TPM \leq \gamma^2 Q. \tag{3.3}$$

To see this, first suppose that $L \in \mathbf{R}^{p\times p}$ and $R \in \mathbf{R}^{q\times q}$ are nonsingular and diagonal, and (3.2) holds. Then (3.3) holds with $P = L^T L$ and $Q = (RR^T)^{-1}$. Conversely, suppose that (3.3) holds for diagonal $P \in \mathbf{R}^{p\times p}$ and $Q \in \mathbf{R}^{q\times q}$ with $P > 0$, $Q > 0$. Then, (3.2) holds for $L = P^{1/2}$ and $R = Q^{-1/2}$.

Hence we can solve (3.1) by solving the GEVP:

$$
\begin{array}{ll}
\text{minimize} & \gamma^2 \\
\text{subject to} & P \in \mathbf{R}^{p\times p} \text{ and diagonal}, \qquad P > 0, \\
& Q \in \mathbf{R}^{q\times q} \text{ and diagonal}, \qquad Q > 0 \\
& Q \le M^T P M \le \gamma^2 Q
\end{array}
$$

Remark: This result is readily extended to handle scaling matrices that have a given block-diagonal structure, and more generally, the constraint that one or more blocks are equal. Complex matrices are readily handled by substituting M^* (complex-conjugate transpose) for M^T.

3.2 Minimizing Condition Number of a Positive-Definite Matrix

A related problem is minimizing the condition number of a positive-definite matrix M that depends affinely on the variable x, subject to the LMI constraint $F(x) > 0$. This problem can be reformulated as the GEVP

$$
\begin{array}{ll}
\text{minimize} & \gamma \\
\text{subject to} & F(x) > 0, \qquad \mu > 0, \qquad \mu I < M(x) < \gamma\mu I
\end{array}
\tag{3.4}
$$

(the variables here are x, μ, and γ).

We can reformulate this GEVP as an EVP as follows. Suppose

$$
M(x) = M_0 + \sum_{i=1}^m x_i M_i, \qquad F(x) = F_0 + \sum_{i=1}^m x_i F_i.
$$

Defining the new variables $\nu = 1/\mu$, $\tilde{x} = x/\mu$, we can express (3.4) as the EVP with variables $\nu = 1/\mu$, $\tilde{x} = x/\mu$ and γ:

$$
\begin{array}{ll}
\text{minimize} & \gamma \\
\text{subject to} & \nu F_0 + \sum_{i=1}^m \tilde{x}_i F_i > 0, \qquad I < \nu M_0 + \sum_{i=1}^m \tilde{x}_i M_i < \gamma I
\end{array}
$$

3.3 Minimizing Norm by Scaling

The optimal diagonally scaled norm of a matrix $M \in \mathbf{C}^{n\times n}$ is defined as

$$
\nu(M) \stackrel{\Delta}{=} \inf\left\{ \|DMD^{-1}\| \;\middle|\; D \in \mathbf{C}^{n\times n} \text{ is diagonal and nonsingular} \right\},
$$

where $\|M\|$ is the norm (i.e., largest singular value) of the matrix M. We will show that $\nu(M)$ can be computed by solving a GEVP.

Note that

$$\begin{aligned}\nu(M) &= \inf\left\{\gamma \;\middle|\; \begin{array}{l}\left(DMD^{-1}\right)^*\left(DMD^{-1}\right) < \gamma^2 I \\ \text{for some diagonal, nonsingular } D\end{array}\right\} \\ &= \inf\left\{\gamma \mid M^*D^*DM \le \gamma^2 D^*D \text{ for some diagonal, nonsingular } D\right\} \\ &= \inf\left\{\gamma \mid M^*PM \le \gamma^2 P \text{ for some diagonal } P = P^T > 0\right\}.\end{aligned}$$

Therefore $\nu(M)$ is the optimal value of the GEVP

$$\begin{array}{ll}\text{minimize} & \gamma \\ \text{subject to} & P > 0 \text{ and diagonal}, \qquad M^*PM < \gamma^2 P\end{array}$$

Remark: This result can be extended in many ways. For example, the more general case of block-diagonal similarity-scaling, with equality constraints among the blocks, is readily handled. As another example, we can solve the problem of simultaneous optimal diagonal scaling of several matrices $M_1, \ldots, M_p$: To compute

$$\inf\left\{\max_{i=1,\ldots,p} \left\|DM_iD^{-1}\right\| \;\middle|\; \begin{array}{l} D \in \mathbf{C}^{n\times n} \text{ is diagonal} \\ \text{and nonsingular}\end{array}\right\},$$

we simply compute the optimal value of the GEVP

$$\begin{array}{ll}\text{minimize} & \gamma \\ \text{subject to} & P > 0 \text{ and diagonal}, \qquad M_i^*PM_i < \gamma^2 P, \quad i = 1, \ldots, p\end{array}$$

Another closely related quantity, which plays an important role in the stability analysis of linear systems with uncertain parameters, is

$$\inf\left\{\gamma \;\middle|\; \begin{array}{l} P > 0, \quad M^*PM + i(M^*G - GM) < \gamma^2 P, \\ P \text{ and } G \text{ diagonal and real}\end{array}\right\}, \tag{3.5}$$

which can also be computed by solving a GEVP.

3.4 Rescaling a Matrix Positive-Definite

We are given a matrix $M \in \mathbf{C}^{n\times n}$, and ask whether there is a diagonal matrix $D > 0$ such that the Hermitian part of DM is positive-definite. This is true if and only if the LMI

$$M^*D + DM > 0, \qquad D > 0 \tag{3.6}$$

is feasible. So determining whether a matrix can be rescaled positive-definite is an LMIP.

When this LMI is feasible, we can find a feasible scaling D with the smallest condition number by solving the EVP

$$\begin{array}{ll}\text{minimize} & \mu \\ \text{subject to} & D \text{ diagonal}, \qquad M^*D + DM > 0, \qquad I < D < \mu^2 I\end{array}$$

Remark: The LMI (3.6) can also be interpreted as the condition for the existence of a diagonal quadratic Lyapunov function that proves stability of $-M$; see the Notes and References and §10.3.

3.5 Matrix Completion Problems

In a *matrix completion problem* some entries of a matrix are fixed and the others are to be chosen so the resulting ("completed") matrix has some property, e.g., is Toeplitz and positive-definite. Many of these problems can be cast as LMIPs. The simplest example is positive-definite completion: Some entries of a symmetric matrix are given, and we are to complete the matrix to make it positive-definite. Evidently this is an LMIP.

We can easily recover some known results about this problem. We can assume that the diagonal entries are specified (otherwise we simply make them very large, i.e., eliminate them; see §2.6.1). Then the set of all positive-definite completions, if nonempty, is bounded. Hence the associated LMI has an analytic center $A^\star$ (see §2.4.1). The matrix $A^\star$ is characterized by $\mathbf{Tr}\, A^{\star -1} B = 0$ for every $B = B^T$ that has zero entries where A has specified entries (see §2.4.2). Thus the matrix $A^{\star -1}$ has a zero entry in every location corresponding to an unspecified entry in the original matrix, i.e., it has the same sparsity structure as the original matrix. Depending on the pattern of specified entries, this condition can lead to an "analytic solution" of the problem. But of course an arbitrary positive-definite completion problem is readily solved as an LMIP.

Let us mention a few simple generalizations of this matrix completion problem that have not been considered in the literature but are readily solved as LMIPs, and might be useful in applications.

Suppose that for each entry of the matrix we specify an upper and lower bound, in other words, we consider an "interval matrix". For entries that are completely specified, the lower and upper bound coincide; for entries that are completely free the lower bound is $-\infty$ and the upper bound is $+\infty$. The problem of completing the matrix, subject to the bounds on each entry, to make it positive-definite (or a contraction, i.e., have norm less than one) is readily formulated as an LMIP.

As another example, suppose we are given a set of matrices that have some given sparsity pattern. We want to determine a matrix with the complementary sparsity pattern that, when added to each of the given matrices, results in a positive-definite (or contractive, etc.) matrix. We might call this the "simultaneous matrix completion problem" since we seek a completion that serves for multiple original matrices. This problem is also readily cast as an LMIP.

As a more specific example, suppose we are given upper and lower bounds for some of the correlation coefficients of a stationary time-series, and must determine some complementary correlation coefficients that are consistent with them. In other words, we have a Toeplitz matrix with some entries within given intervals, while the others are completely free (unspecified). We must determine these free entries so that the Toeplitz matrix is positive-definite, for any choice of the specified elements in the intervals. This is an LMIP.

3.6 Quadratic Approximation of a Polytopic Norm

Consider a piecewise linear norm $\|\cdot\|_{\rm pl}$ on $\mathbf{R}^n$ defined by

$$\|z\|_{\rm pl} \stackrel{\Delta}{=} \max_{i=1,\ldots,p} \left|a_i^T z\right|,$$

where $a_i \in \mathbf{R}^n$ for $i = 1, \ldots, p$.

For any $P > 0$, the quadratic norm defined by $\|z\|_P \stackrel{\Delta}{=} \sqrt{z^T P z} = \|P^{1/2} z\|$ satisfies

$$1/\sqrt{\alpha}\|z\|_P \leq \|z\|_{\rm pl} \leq \sqrt{\alpha}\|z\|_P \text{ for all } z,$$

for some constant $\alpha \geq 1$. Thus, the quadratic norm $\sqrt{z^T P z}$ approximates $\|z\|_{\rm pl}$ within a factor of α. We will show that the problem of finding P that minimizes α, i.e., the problem of determining the optimal quadratic norm approximation of $\|\cdot\|_{\rm pl}$, is an EVP.

Let $v_1, \ldots, v_L$ be the vertices of the unit ball $\mathcal{B}_{\rm pl}$ of $\|\cdot\|_{\rm pl}$, so that

$$\mathcal{B}_{\rm pl} \stackrel{\Delta}{=} \{z \mid \|z\|_{\rm pl} \leq 1\} = \mathbf{Co}\{v_1, \ldots, v_L\}$$

and let $\mathcal{B}_P$ denote the unit ball of $\|\cdot\|_P$. Then

$$1/\sqrt{\alpha}\|z\|_P \leq \|z\|_{\rm pl} \leq \sqrt{\alpha}\|z\|_P \tag{3.7}$$

is equivalent to

$$1/\sqrt{\alpha}\mathcal{B}_P \subseteq \mathcal{B}_{\rm pl} \subseteq \sqrt{\alpha}\mathcal{B}_P.$$

The first inclusion, $1/\sqrt{\alpha}\mathcal{B}_P \subseteq \mathcal{B}_{\rm pl}$, is equivalent to

$$a_i^T P^{-1} a_i \leq \alpha, \quad i = 1, \ldots, p,$$

and the second, $\mathcal{B}_{\rm pl} \subseteq \sqrt{\alpha}\mathcal{B}_P$, is equivalent to

$$v_i^T P v_i \leq \alpha, \quad i = 1, \ldots, L.$$

Therefore we can minimize α such that (3.7) holds for some $P > 0$ by solving the EVP

$$\begin{array}{ll} \text{minimize} & \alpha \\ \text{subject to} & v_i^T P v_i \leq \alpha, \quad i = 1, \ldots, L, \qquad \begin{bmatrix} P & a_i \\ a_i^T & \alpha \end{bmatrix} \geq 0, \quad i = 1, \ldots, p \end{array}$$

Remark: The number of vertices L of the unit ball of $\|\cdot\|_{\rm pl}$ can grow exponentially in p and n. Therefore the result is uninteresting from the point of view of computational complexity and of limited practical use except for problems with low dimensions.

The optimal factor α in this problem never exceeds $\sqrt{n}$. Indeed, by computing the maximum volume ellipsoid that lies inside the unit ball of $\|\cdot\|_{\rm pl}$, (which does not require us to find the vertices v_i, $i = 1, \ldots, L$), we can find (in polynomial-time) a norm for which α is less than $\sqrt{n}$. See the Notes and References for more details.

3.7 Ellipsoidal Approximation

The problem of approximating some subset of $\mathbf{R}^n$ with an ellipsoid arises in many fields and has a long history; see the Notes and References. To pose such a problem precisely we need to know how the subset is described, whether we seek an inner or outer approximation, and how the approximation will be judged (volume, major or minor semi-axis, etc.).

In some cases the problem can be cast as a CP or an EVP, and hence solved exactly. As an example consider the problem of finding the ellipsoid centered around the origin of smallest volume that contains a polytope described by its vertices. We saw in §2.2.4 that this problem can be cast as a CP.

In other cases, we can compute an approximation of the optimal ellipsoid by solving a CP or an EVP. In some of these cases, the problem is known to be NP-hard, so it is unlikely that the problem can be reduced (polynomially) to an LMI problem. It also suggests that the approximations obtained by solving a CP or EVP will not be good for *all* instances of problem data, although the approximations may be good on "typical" problems, and hence of some use in practice.

As an example consider the problem of finding the ellipsoid of smallest volume that contains a polytope described by a set of linear inequalities, i.e., $\{ x \mid a_i^T x \leq b_i,\ i = 1, \ldots, p \}$. (This is the same problem as described above, but with a different description of the polytope.) This problem is NP-hard. Indeed, consider the problem of simply verifying that a given, fixed ellipsoid contains a polytope described by a set of linear inequalities. This problem is equivalent to the general concave quadratic programming problem, which is NP-complete.

In this section we consider subsets formed from ellipsoids $\mathcal{E}_1, \ldots, \mathcal{E}_p$ in various ways: union, intersection, and addition. We approximate these sets by an ellipsoid $\mathcal{E}_0$.

We describe an ellipsoid $\mathcal{E}$ in two different ways. The first description uses convex quadratic functions:

$$\mathcal{E} = \{x \mid T(x) \leq 0, \}\,, \qquad T(x) = x^T A x + 2x^T b + c, \tag{3.8}$$

where $A = A^T > 0$ and $b^T A^{-1} b - c > 0$ (which ensures that $\mathcal{E}$ is nonempty and does not reduce to a single point). Note that this description is homogeneous, i.e., we can scale A, b and c by any positive factor without affecting $\mathcal{E}$.

The volume of $\mathcal{E}$ is given by

$$\mathbf{vol}(\mathcal{E})^2 = \beta \det\left((b^T A^{-1} b - c) A^{-1}\right)$$

where β is a constant that depends only on the dimension of x, i.e., n. The diameter of $\mathcal{E}$ (i.e., twice the maximum semi-axis length) is

$$2\sqrt{(b^T A^{-1} b - c)\lambda_{\max}(A^{-1})}.$$

We will also describe an ellipsoid as the image of the unit ball under an affine mapping with symmetric positive-definite matrix:

$$\mathcal{E} = \{ x \mid (x - x_c)^T P^{-2} (x - x_c) \leq 1 \} = \{ Pz + x_c \mid \|z\| \leq 1 \}, \tag{3.9}$$

where $P = P^T > 0$. (This representation is unique, i.e., P and x_c are uniquely determined by the ellipsoid $\mathcal{E}$.) The volume of $\mathcal{E}$ is then proportional to $\det P$, and its diameter is $2\lambda_{\max}(P)$.

Each of these descriptions of $\mathcal{E}$ is readily converted into the other. Given a representation (3.8) (i.e., A, b, and c) we form the representation (3.9) with

$$P = \sqrt{b^T A^{-1} b - c}\, A^{-1/2}, \qquad x_c = -A^{-1} b.$$

Given the representation (3.9), we can form a representation (3.8) of $\mathcal{E}$ with

$$A = P^{-2}, \qquad b = -P^{-2}x_c, \qquad c = x_c^T P^{-2} x_c - 1.$$

(This forms one representation of $\mathcal{E}$; we can form every representation of $\mathcal{E}$ by scaling these A, b, c by positive factors.)

3.7.1 Outer approximation of union of ellipsoids

We seek a small ellipsoid $\mathcal{E}_0$ that covers the union of ellipsoids $\mathcal{E}_1, \ldots, \mathcal{E}_p$ (or equivalently, the convex hull of the union, i.e., $\mathbf{Co}\bigcup_{i=1}^p \mathcal{E}_i$). We will describe these ellipsoids via the associated quadratic functions $T_i(x) = x^T A_i x + 2x^T b_i + c_i$. We have

$$\mathcal{E}_0 \supseteq \bigcup_{i=1}^p \mathcal{E}_i \tag{3.10}$$

if and only if $\mathcal{E}_i \subseteq \mathcal{E}_0$ for $i = 1, \ldots, p$. This is true if and only if, for each i, every x such that $T_i(x) \le 0$ satisfies $T_0(x) \le 0$. By the $\mathcal{S}$-procedure, this is true if and only if there exist nonnegative scalars $\tau_1, \ldots, \tau_p$ such that

$$\text{for every } x, \qquad T_0(x) - \tau_i T_i(x) \le 0, \quad i = 1, \ldots, p,$$

or, equivalently, such that

$$\begin{bmatrix} A_0 & b_0 \\ b_0^T & c_0 \end{bmatrix} - \tau_i \begin{bmatrix} A_i & b_i \\ b_i^T & c_i \end{bmatrix} \le 0, \quad i = 1, \ldots, p.$$

Since our representation of $\mathcal{E}_0$ is homogeneous, we will now normalize A_0, b_0 and c_0 in a convenient way: such that $b_0^T A_0^{-1} b_0 - c_0 = 1$. In other words we set

$$c_0 = b_0^T A_0^{-1} b_0 - 1 \tag{3.11}$$

and parametrize $\mathcal{E}_0$ by A_0 and b_0 alone. Thus our condition becomes:

$$\begin{bmatrix} A_0 & b_0 \\ b_0^T & b_0^T A_0^{-1} b_0 - 1 \end{bmatrix} - \tau_i \begin{bmatrix} A_i & b_i \\ b_i^T & c_i \end{bmatrix} \le 0, \quad i = 1, \ldots, p.$$

Using Schur complements, we obtain the equivalent LMI

$$\begin{bmatrix} A_0 & b_0 & 0 \\ b_0^T & -1 & b_0^T \\ 0 & b_0 & -A_0 \end{bmatrix} - \tau_i \begin{bmatrix} A_i & b_i & 0 \\ b_i^T & c_i & 0 \\ 0 & 0 & 0 \end{bmatrix} \le 0, \quad i = 1, \ldots, p, \tag{3.12}$$

with variables A_0, b_0, and τ_1, ..., τ_p. To summarize, the condition (3.10) is equivalent to the existence of nonnegative $\tau_1, \ldots, \tau_p$ such that (3.12) holds. (Since $A_0 > 0$, it follows that the τ_i must, in fact, be positive.)

With the normalization (3.11), the volume of $\mathcal{E}_0$ is proportional to $\sqrt{\det A_0^{-1}}$. Thus we can find the smallest volume ellipsoid containing the union of ellipsoids $\mathcal{E}_1$, ..., $\mathcal{E}_p$ by solving the CP (with variables A_0, b_0, and τ_1, ..., τ_p)

$$\begin{array}{ll} \text{minimize} & \log\det A_0^{-1} \\ \text{subject to} & A_0 > 0, \qquad \tau_1 \ge 0, \ldots, \tau_p \ge 0, \qquad (3.12) \end{array}$$

This ellipsoid is called the *Löwner–John* ellipsoid for $\bigcup_{i=1}^{p} \mathcal{E}_i$, and satisfies several nice properties (e.g., affine invariance, certain bounds on how well it approximates $\mathbf{Co} \bigcup \mathcal{E}_i$); see the Notes and References.

We can find the ellipsoid (or equivalently, sphere) of smallest diameter that contains $\bigcup \mathcal{E}_i$ by minimizing $\lambda_{\max}(A_0^{-1})$ subject to the constraints of the previous CP, which is equivalent to solving the EVP

$$\begin{array}{ll} \text{maximize} & \lambda \\ \text{subject to} & A_0 > \lambda I, \qquad \tau_1 \geq 0, \ldots, \tau_p \geq 0, \qquad (3.12) \end{array}$$

The optimal diameter is $2/\sqrt{\lambda^{\text{opt}}}$, where λ^{opt} is an optimal value of the EVP above.

3.7.2 Outer approximation of the intersection of ellipsoids

Simply verifying that

$$\mathcal{E}_0 \supseteq \bigcap_{i=1}^{p} \mathcal{E}_i \tag{3.13}$$

holds, given $\mathcal{E}_0$, $\mathcal{E}_1$, ..., $\mathcal{E}_p$, is NP-complete, so we do not expect to recast it as one of our LMI problems. A fortiori, we do not expect to cast the problem of finding the smallest volume ellipsoid covering the intersection of some ellipsoids as a CP.

We will describe three ways that we can compute suboptimal ellipsoids for this problem by solving CPs. The first method uses the $\mathcal{S}$-procedure to derive an LMI that is a sufficient condition for (3.13) to hold.

Once again we normalize our representation of $\mathcal{E}_0$ so that (3.11) holds. From the $\mathcal{S}$-procedure we obtain the condition: there exist positive scalars $\tau_1, \ldots, \tau_p$ such that

$$\begin{bmatrix} A_0 & b_0 \\ b_0^T & b_0^T A_0^{-1} b_0 - 1 \end{bmatrix} - \sum_{i=1}^{p} \tau_i \begin{bmatrix} A_i & b_i \\ b_i^T & c_i \end{bmatrix} \leq 0,$$

which can be written as the LMI (in variables A_0, b_0, and $\tau_1, \ldots, \tau_p$):

$$\begin{bmatrix} A_0 & b_0 & 0 \\ b_0^T & -1 & b_0^T \\ 0 & b_0 & -A_0 \end{bmatrix} - \sum_{i=1}^{p} \tau_i \begin{bmatrix} A_i & b_i & 0 \\ b_i^T & c_i & 0 \\ 0 & 0 & 0 \end{bmatrix} \leq 0. \tag{3.14}$$

This LMI is *sufficient* but not necessary for (3.13) to hold, i.e., it characterizes some but not all of the ellipsoids that cover the intersection of $\mathcal{E}_1$, ..., $\mathcal{E}_p$. We can find the best such outer approximation (i.e., the one of smallest volume) by solving the CP (with variables A_0, b_0, and $\tau_1, \ldots, \tau_p$)

$$\begin{array}{ll} \text{minimize} & \log\det A_0^{-1} \\ \text{subject to} & A_0 > 0, \qquad \tau_1 \geq 0, \ldots, \tau_p \geq 0, \qquad (3.14) \end{array} \tag{3.15}$$

Remark: The intersection $\bigcap_{i=1}^{p} \mathcal{E}_i$ is empty or a single point if and only if there exist nonnegative $\tau_1, \ldots, \tau_p$, not all zero, such that

$$\sum_{i=1}^{p} \tau_i \begin{bmatrix} A_i & b_i \\ b_i^T & c_i \end{bmatrix} \geq 0.$$

In this case (3.15) is unbounded below.

Another method that can be used to find a suboptimal solution for the problem of determining the minimum volume ellipsoid that contains the intersection of $\mathcal{E}_1, \ldots, \mathcal{E}_p$ is to first compute the *maximum* volume ellipsoid that is *contained in* the intersection, which can be cast as a CP (see the next section). If this ellipsoid is scaled by a factor of n about it center then it is guaranteed to *contain* the intersection. See the Notes and References for more discussion.

Yet another method for producing a suboptimal solution is based on the idea of the analytic center (see §2.4.1). Suppose that $\mathcal{E}_1, \ldots, \mathcal{E}_p$ are given in the form (3.9), with $x_1, \ldots, x_p$ and $P_1, \ldots, P_p$. The LMI

$$F(x) = \mathbf{diag}\left(\begin{bmatrix} I & P_1^{-1}(x-x_1) \\ (x-x_1)^T P_1^{-1} & 1 \end{bmatrix}, \ldots, \begin{bmatrix} I & P_p^{-1}(x-x_p) \\ (x-x_p)^T P_p^{-1} & 1 \end{bmatrix}\right) > 0$$

in the variable x has as feasible set the interior of $\bigcap_{i=1}^p \mathcal{E}_i$. We assume now that the intersection $\bigcap_{i=1}^p \mathcal{E}_i$ has nonempty interior (i.e., is nonempty and not a single point). Let $x^\star$ denote its analytic center, i.e.,

$$x^\star = \underset{F(x)>0}{\arg\min} \quad \log\det F(x)^{-1}$$

and let H denote the Hessian of $\log\det F(x)^{-1}$ at $x^\star$. Then it can be shown that

$$\mathcal{E}_0 \stackrel{\Delta}{=} \left\{x \mid (x-x^\star)^T H (x-x^\star) \le 1\right\} \subseteq \bigcap_{i=1}^p \mathcal{E}_i.$$

In fact we can give a bound on how poorly $\mathcal{E}_0$ approximates the intersection. It can be shown that

$$\left\{x \mid (x-x^\star)^T H (x-x^\star) \le (3p+1)^2\right\} \supseteq \bigcap_{i=1}^p \mathcal{E}_i.$$

Thus, when $\mathcal{E}_0$ is enlarged by a factor of $3p+1$, it covers $\bigcap_{i=1}^p \mathcal{E}_i$. See the Notes and References for more discussion of this.

3.7.3 Inner approximation of the intersection of ellipsoids

The ellipsoid $\mathcal{E}_0$ is contained in the intersection of the ellipsoids $\mathcal{E}_1, \ldots, \mathcal{E}_p$ if and only if for every x satisfying

$$(x-x_0)^T P_0^{-2}(x-x_0) \le 1,$$

we have

$$(x-x_i)^T P_i^{-2}(x-x_i) \le 1, \quad i=1,\ldots,p.$$

Using the $\mathcal{S}$-procedure, this is equivalent to the existence of nonnegative $\lambda_1, \ldots, \lambda_p$ satisfying

$$\begin{array}{ll} \text{for every } x, & (x-x_i)^T P_i^{-2}(x-x_i) \\ & -\lambda_i (x-x_0)^T P_0^{-2}(x-x_0) \le 1-\lambda_i, \quad i=1,\ldots,p. \end{array} \tag{3.16}$$

We prove in the Notes and References section that (3.16) is equivalent to

$$\begin{bmatrix} -P_i^2 & x_i - x_0 & P_0 \\ (x_i - x_0)^T & \lambda_i - 1 & 0 \\ P_0 & 0 & -\lambda_i I \end{bmatrix} \leq 0, \quad i = 1, \ldots, p. \tag{3.17}$$

Therefore, we obtain the largest volume ellipsoid contained in the intersection of the ellipsoids $\mathcal{E}_1, \ldots, \mathcal{E}_p$ by solving the CP (with variables P_0, x_0, $\lambda_1, \ldots, \lambda_p$)

$$\begin{array}{ll} \text{minimize} & \log\det P_0^{-1} \\ \text{subject to} & P_0 > 0, \qquad \lambda_1 \geq 0, \ldots, \lambda_p \geq 0, \qquad (3.17) \end{array}$$

Among the ellipsoids contained in the intersection of $\mathcal{E}_1, \ldots, \mathcal{E}_p$, we can find one that maximizes the smallest semi-axis by solving

$$\begin{array}{ll} \text{minimize} & \lambda \\ \text{subject to} & \lambda I > P_0 > 0, \qquad \lambda_1 \geq 0, \ldots, \lambda_p \geq 0, \qquad (3.17) \end{array}$$

Equivalently, this yields the sphere of largest diameter contained in the intersection.

3.7.4 Outer approximation of the sum of ellipsoids

The sum of the ellipsoids $\mathcal{E}_1, \ldots, \mathcal{E}_p$ is defined as

$$\mathcal{A} \stackrel{\Delta}{=} \{x_1 + \cdots + x_p \mid x_1 \in \mathcal{E}_1, \ldots, x_p \in \mathcal{E}_p \,.\}$$

We seek a small ellipsoid $\mathcal{E}_0$ containing $\mathcal{A}$.

The ellipsoid $\mathcal{E}_0$ contains $\mathcal{A}$ if and only if for every $x_1, \ldots, x_p$ such that $T_i(x_i) \leq 0$, $i = 1, \ldots, p$, we have

$$T_0\left(\sum_{i=1}^p x_i\right) \leq 0.$$

By application of the $\mathcal{S}$-procedure, this condition is true if

$$\begin{array}{l} \text{there exist } \tau_i \geq 0,\ i = 1, \ldots, p \text{ such that} \\ \text{for every } x_i, \quad i = 1, \ldots, p, \qquad T_0\left(\sum_{i=1}^p x_i\right) - \sum_{i=1}^p \tau_i T_i(x_i) \leq 0. \end{array} \tag{3.18}$$

Once again we normalize our representation of $\mathcal{E}_0$ so that (3.11) holds. Let x denote the vector made by stacking $x_1, \ldots, x_p$, i.e.,

$$x = \begin{bmatrix} x_1 \\ \vdots \\ x_p \end{bmatrix}.$$

Then both $T_0\left(\sum_{i=1}^p x_i\right)$ and $\sum_{i=1}^p \tau_i T_i(x_i)$ are quadratic functions of x. Indeed, let E_i denote the matrix such that $x_i = E_i x$, and define

$$E_0 = \sum_{i=1}^p E_i, \qquad \tilde{A}_i = E_i^T A_i E_i, \qquad \tilde{b}_i = E_i^T b_i, \quad i = 0, \ldots, p.$$

With this notation, we see that condition (3.18) is equivalent to

there exist $\tau_i \geq 0$, $i = 1, \ldots, p$ such that

$$\begin{bmatrix} \tilde{A}_0 & \tilde{b}_0 \\ \tilde{b}_0^T & b_0^T A_0^{-1} b_0 - 1 \end{bmatrix} - \sum_{i=1}^{p} \tau_i \begin{bmatrix} \tilde{A}_i & \tilde{b}_i \\ \tilde{b}_i^T & c_i \end{bmatrix} \leq 0. \tag{3.19}$$

This condition is readily written as an LMI in variables A_0, b_0, $\tau_1, \ldots, \tau_p$ using Schur complements:

$$\begin{bmatrix} \tilde{A}_0 & \tilde{b}_0 & 0 \\ \tilde{b}_0^T & -1 & b_0^T \\ 0 & b_0 & -A_0 \end{bmatrix} - \sum_{i=1}^{p} \tau_i \begin{bmatrix} \tilde{A}_i & \tilde{b}_i & 0 \\ \tilde{b}_i^T & c_i & 0 \\ 0 & 0 & 0 \end{bmatrix} \leq 0. \tag{3.20}$$

Therefore, we compute the minimum volume ellipsoid, proven to contain the sum of $\mathcal{E}_1, \ldots, \mathcal{E}_p$ via the $\mathcal{S}$-procedure, by minimizing $\log\det A_0^{-1}$ subject to $A_0 > 0$ and (3.20). This is a CP.

Notes and References

Matrix scaling problems

The problem of scaling a matrix to reduce its condition number is considered in Forsythe and Straus [FS55] (who describe conditions for the scaling T to be optimal in terms of a new variable $R = T^{-*}T^{-1}$, just as we do). Other references on the problem of improving the condition number of a matrix via diagonal or block-diagonal scaling are [BAU63, GV74, SHA85, WAT91]. We should point out that the results of §3.1 are not practically useful in numerical analysis, since the cost of computing the optimal scalings (via the GEVP) exceeds the cost of solving the problem for which the scalings are required (e.g., solving a set of linear equations).

A special case of the problem of minimizing the condition number of a positive-definite matrix is considered in [EHM92, EHM93]. Finally, Khachiyan and Kalantari ([KK92]) consider the problem of diagonally scaling a positive semidefinite matrix via interior-point methods and give a complete complexity analysis for the problem.

The problem of similarity-scaling a matrix to minimize its norm appears often in control applications (see for example [DOY82, SAF82]). A recent survey article on this topic is by Packard and Doyle [PD93]. A related quantity is the one-sided multivariable stability margin, described in [TSC92]; the problem of computing this quantity can be easily transformed into a GEVP. The quantity (3.5), used in the analysis of linear systems with parameter uncertainties, was introduced by Fan, Tits and Doyle in [FTD91] (see also [BFBE92]). Some closely related quantities were considered in [SC93, SL93A, CS92B]; see also Chapter 8 in this book, where a different approach to the same problem and its extensions is considered. We also mention some other related quantities found in the articles by Kouvaritakis and Latchman [KL85A, KL85B, LN92], and Rotea and Prasanth [RP93, RP94], which can be computed via GEVPs.

Diagonal quadratic Lyapunov functions

The problem of rescaling a matrix M with a diagonal matrix $D > 0$ such that its Hermitian part is positive-definite is considered in the articles by Barker et al. [BBP78] and Khalil [KHA82]. The problem is the same as determining the existence of diagonal quadratic Lyapunov function for a given linear system; see [GER85, HU87, BY89, KRA91]. In [KB93], Kaszkurewicz and Bhaya give an excellent survey on the topic of diagonal Lyapunov functions, including many references and examples from mathematical ecology, power systems, and digital filter stability. Diagonal Lyapunov functions arise in the analysis of positive orthant stability; see §10.3.

Matrix completion problems

Matrix completions problems are addressed in the article of Dym and Gohberg [DG81], in which the authors examine conditions for the existence of completions of band matrices such that the inverse is also a band matrix. In [GJSW84], their results are extended to a more general class of partially specified matrices. In both papers the "sparsity pattern" result mentioned in §3.5 appears. In [BJL89], Barrett et al. provide an algorithm for finding the completion maximizing the determinant of the completed matrix (i.e., the analytic center $A^\star$ mentioned in §3.5). In [HW93], Helton and Woerdeman consider the problem of minimum norm extension of Hankel matrices. In [GFS93], Grigoriadis, Frazho and Skelton consider the problem of approximating a given symmetric matrix by a Toeplitz matrix and propose a solution to this problem via convex optimization. See also [OSA93].

Dancis ([DAN92, DAN93] and Johnson and Lunquist [JL92]) study completion problems in which the rank or inertia of the completed matrix is specified. As far as we know, such problems cannot be cast as LMIPs.

A classic paper on the *contractive completion problem* (i.e., completing a matrix so its norm is less than one) is Davis, Kahan, and Weinberger in [DKW82], in which an analytic solution is given for contractive completion problems with a special block form. The result in this paper is closely related to (and also more general than) the matrix elimination procedure given in §2.6.2. See [WOE90, NW92, BW92] for more references on contractive completion problems.

Maximum volume ellipsoid contained in a symmetric polytope

In this section, we establish some properties of the maximum volume ellipsoid contained in a symmetric polytope described as:

$$\mathcal{P} = \left\{ z \;\middle|\; \left|a_i^T z\right| \le 1, \quad i = 1,\ldots,p \right\}.$$

Consider an ellipsoid described by $\mathcal{E} = \{x \mid x^T Q^{-1} x \le 1\}$ where $Q = Q^T > 0$. Its volume is proportional to $\sqrt{\det Q}$. The condition $\mathcal{E} \subseteq \mathcal{P}$ can be expressed $a_i^T Q a_i \le 1$ for $i = 1,\ldots,p$. Hence we obtain the maximum volume ellipsoid contained in $\mathcal{P}$ by solving the CP

$$\begin{array}{ll} \text{minimize} & \log\det Q^{-1} \\ \text{subject to} & Q > 0, \qquad a_i^T Q a_i \le 1, \quad i = 1,\ldots,p \end{array}$$

Suppose now that Q is optimal. From the standard optimality conditions, there are nonnegative $\lambda_1,\ldots,\lambda_p$ such that

$$Q^{-1} = \sum_{i=1}^{p} \frac{\lambda_i}{a_i^T Q a_i} a_i a_i^T,$$

and $\lambda_i = 0$ if $a_i^T Q a_i = 1$. Multiplying by $Q^{1/2}$ on the left and right and taking the trace yields $\sum_{i=1}^{p} \lambda_i = n$.

For any $x \in \mathbf{R}^n$, we have

$$x^T Q^{-1} x = \sum_{i=1}^{p} \lambda_i \frac{(x^T a_i)^2}{a_i^T Q a_i}.$$

For each i, either $\lambda_i = 0$ or $a_i^T Q a_i = 1$, so we conclude

$$x^T Q^{-1} x = \sum_{i=1}^{p} \lambda_i (x^T a_i)^2.$$

Suppose now that x belongs to $\mathcal{P}$. Then

$$x^T Q^{-1} x \le \sum_{i=1}^{p} \lambda_i = n.$$

Equivalently, $\mathcal{P} \subseteq \sqrt{n}\mathcal{E}$, so we have:

$$\mathcal{E} \subseteq \mathcal{P} \subseteq \sqrt{n}\mathcal{E},$$

i.e., the maximum volume ellipsoid contained in $\mathcal{P}$ approximates $\mathcal{P}$ within the scale factor $\sqrt{n}$.

Ellipsoidal approximation

The problem of computing the minimum volume ellipsoid in $\mathbf{R}^n$ that contains a given convex set was considered by John and Löwner, who showed that the optimal ellipsoid, when shrunk about its center by a factor of n, is contained in the given set (see [JOH85, GLS88]). Such ellipsoids are called Löwner–John ellipsoids. In the case of symmetric sets, the factor n can be improved to $\sqrt{n}$ (using an argument similar to the one in the Notes and References section above). A closely related problem is finding the maximum volume ellipsoid contained in a convex set. Similar results hold for these ellipsoids.

Special techniques have been developed to compute smallest spheres and ellipsoids containing a given set of points in spaces of low dimension; see e.g., Preparata and Shamos [PS85, P.255] or Post [POS84].

Minimum volume ellipsoids containing a given set arise often in control theory. Schweppe and Schlaepfer [SCH68, SS72] use ellipsoids containing the sum and the intersection of two given ellipsoids in the problem of state estimation with norm-bounded disturbances. See also the articles by Bertsekas and Rhodes [BR71], Chernousko [CHE80A, CHE80B, CHE80C] and Kahan [KAH68]. In [RB92], ellipsoidal approximations of robot linkages and workspace are used to rapidly detect or avoid collisions.

Minimum volume ellipsoids have been widely used in system identification; see the articles by Fogel [FOG79], Fogel and Huang [FH82], Deller [DEL89], Belforte, Bona and Cerone [BBC90], Lau, Kosut and Boyd [LKB90, KLB90, KLB92], Cheung, Yurkovich and Passino [CYP93] and the book by Norton [NOR86].

In the ellipsoid algorithm we encountered the minimum volume ellipsoid that contains a given half-ellipsoid; the references on the ellipsoid algorithm describe various extensions.

Efficient interior-point methods for computing minimum volume outer and maximum volume inner ellipsoids for various sets are described in Nesterov and Nemirovskii [NN94]. See also [KT93].

Ellipsoidal approximation via analytic centers

Ellipsoidal approximations via barrier functions arise in the theory of interior-point algorithms, and can be traced back to Dikin [DIK67] and Karmarkar [KAR84] for polytopes described by a set of linear inequalities. For the case of more general LMIs, see Nesterov and Nemirovskii [NN94], Boyd and El Ghaoui [BE93], and Jarre [JAR93C].

These approximations have the property that they are easy to compute, and come with provable bounds on how suboptimal they are. These bounds have the following form: if the inner ellipsoid is stretched about its center by some factor α (which depends only on the dimension of the space, the type and size of the constraints) then the resulting ellipsoid covers the feasible set. The papers cited above give *different* values of α; one may be better than another, depending on the number and type of constraints.

In contrast, the maximum volume inner ellipsoid, i.e., the Löwner–John ellipsoid, is more difficult to compute (still polynomial, however) but the factor α depends only on the dimension of the space and not on the type or number of constraints, and is always smaller than the ones obtained via barrier functions. So an ellipsoidal approximation from a barrier function can serve as a "cheap substitute" for the Löwner–John ellipsoid.

Proof of lemma in §3.7.3

We prove that the statement

$$\text{for every } x, \qquad (x-x_1)^T P_1^{-2}(x-x_1) - \lambda(x-x_0)^T P_0^{-2}(x-x_0) \leq 1-\lambda \tag{3.21}$$

where P_0, P_1 are positive matrices, $\lambda \geq 0$, is equivalent to the matrix inequality

$$\begin{bmatrix} -P_1^2 & x_1-x_0 & P_0 \\ (x_1-x_0)^T & \lambda-1 & 0 \\ P_0 & 0 & -\lambda I \end{bmatrix} \leq 0. \tag{3.22}$$

The maximum over x of

$$(x-x_1)^T P_1^{-2}(x-x_1) - \lambda(x-x_0)^T P_0^{-2}(x-x_0) \tag{3.23}$$

is finite if and only if $P_0^2 - \lambda P_1^2 \leq 0$ (which implies that $\lambda > 0$) and there exists x^* satisfying

$$P_1^{-2}(x^*-x_1) = \lambda P_0^{-2}(x^*-x_0), \tag{3.24}$$

or, equivalently, satisfying

$$(P_0^2/\lambda - P_1^2)P_1^{-2}(x^*-x_1) = x_1 - x_0. \tag{3.25}$$

In this case, x^* is a maximizer of (3.23). Using (3.24), condition (3.21) implies

$$(x^*-x_1)^T P_1^{-2}(P_1^2 - P_0^2/\lambda)P_1^{-2}(x^*-x_1) \leq 1-\lambda.$$

Using Schur complements, this last condition is equivalent to

$$\begin{bmatrix} \lambda-1 & (x^*-x_1)^T P_1^{-2}(P_0^2/\lambda - P_1^2) \\ (P_0^2/\lambda - P_1^2)P_1^{-2}(x^*-x_1) & P_0^2/\lambda - P_1^2 \end{bmatrix} \leq 0.$$

Using (3.25), this implies

$$\begin{bmatrix} \lambda-1 & (x_1-x_0)^T \\ x_1-x_0 & P_0^2/\lambda - P_1^2 \end{bmatrix} \leq 0,$$

which is equivalent to (3.22). Conversely, it is easy to show that (3.22) implies (3.21).

Chapter 4

Linear Differential Inclusions

4.1 Differential Inclusions

A differential inclusion (DI) is described by:

$$\dot{x} \in F(x(t), t), \quad x(0) = x_0, \tag{4.1}$$

where F is a set-valued function on $\mathbf{R}^n \times \mathbf{R}_+$. Any $x : \mathbf{R}_+ \to \mathbf{R}^n$ that satisfies (4.1) is called a *solution* or *trajectory* of the DI (4.1). Of course, there can be many solutions of the DI (4.1). Our goal is to establish that various properties are satisfied by all solutions of a given DI. For example, we might show that every trajectory of a given DI converges to zero as $t \to \infty$.

By a standard result called the Relaxation Theorem, we may as well assume $F(x, t)$ is a convex set for every x and t. The DI given by

$$\dot{x} \in \mathbf{Co}\, F(x(t), t), \quad x(0) = x_0 \tag{4.2}$$

is called the *relaxed version* of the DI (4.1). Since $\mathbf{Co}\, F(x(t), t) \supseteq F(x(t), t)$, every trajectory of the DI (4.1) is also a trajectory of relaxed DI (4.2). Very roughly speaking, the Relaxation Theorem states that for many purposes the converse is true. (See the References for precise statements.) As a specific and simple example, it can be shown that for every DI we encounter in this book, the reachable or attainable sets of the DI and its relaxed version coincide, i.e., for every $T \geq 0$

$$\{x(T) \mid x \text{ satisfies } (4.1)\} = \{x(T) \mid x \text{ satisfies } (4.2)\}.$$

In fact we will not need the Relaxation Theorem, or rather, we will always get it "for free"—every result we establish in the next two chapters extends immediately to the relaxed version of the problem. The reason is that when a quadratic Lyapunov function is used to establish some property for the DI (4.1), then the same Lyapunov function establishes the property for the relaxed DI (4.2).

4.1.1 Linear differential inclusions

A linear differential inclusion (LDI) is given by

$$\dot{x} \in \Omega x, \quad x(0) = x_0, \tag{4.3}$$

where Ω is a subset of $\mathbf{R}^{n \times n}$. We can interpret the LDI (4.3) as describing a family of linear time-varying systems. Every trajectory of the LDI satisfies

$$\dot{x} = A(t)x, \quad x(0) = x_0, \tag{4.4}$$

for some $A : \mathbf{R}_+ \to \Omega$. Conversely, for any $A : \mathbf{R}_+ \to \Omega$, the solution of (4.4) is a trajectory of the LDI (4.3). In the language of control theory, the LDI (4.3) might be described as an "uncertain time-varying linear system," with the set Ω describing the "uncertainty" in the matrix $A(t)$.

4.1.2 A generalization to systems

We will encounter a generalization of the LDI described above to linear systems with inputs and outputs. We will consider a system described by

$$\begin{aligned} \dot{x} &= A(t)x + B_u(t)u + B_w(t)w, \quad x(0) = x_0, \\ z &= C_z(t)x + D_{zu}(t)u + D_{zw}(t)w \end{aligned} \tag{4.5}$$

where $x : \mathbf{R}_+ \to \mathbf{R}^n$, $u : \mathbf{R}_+ \to \mathbf{R}^{n_u}$, $w : \mathbf{R}_+ \to \mathbf{R}^{n_w}$, $z : \mathbf{R}_+ \to \mathbf{R}^{n_z}$. x is referred to as the state, u is the control input, w is the exogenous input and z is the output. The matrices in (4.5) satisfy

$$\begin{bmatrix} A(t) & B_u(t) & B_w(t) \\ C_z(t) & D_{zu}(t) & D_{zw}(t) \end{bmatrix} \in \Omega. \tag{4.6}$$

for all $t \geq 0$, where $\Omega \subseteq \mathbf{R}^{(n+n_z)\times(n+n_u+n_w)}$. We will be more specific about the form of Ω shortly.

In some applications we can have one or more of the integers n_u, n_w, and n_z equal to zero, which means that the corresponding variable is not used. For example, the LDI $\dot{x} \in \Omega x$ results when $n_u = n_w = n_z = 0$. In order not to introduce another term to describe the set of all solutions of (4.5) and (4.6), we will call (4.5) and (4.6) a *system described by LDIs* or simply, an LDI.

4.2 Some Specific LDIs

We now describe some specific families of LDIs that we will encounter in the next two chapters.

4.2.1 Linear time-invariant systems

When Ω is a singleton, the LDI reduces to the linear time-invariant (LTI) system

$$\begin{aligned} \dot{x} &= Ax + B_u u + B_w w, \quad x(0) = x_0, \\ z &= C_z x + D_{zu} u + D_{zw} w, \end{aligned}$$

where

$$\Omega = \left\{ \begin{bmatrix} A & B_u & B_w \\ C_z & D_{zu} & D_{zw} \end{bmatrix} \right\}. \tag{4.7}$$

Although most of the results of Chapters 5–7 are well-known for LTI systems, some are new; we will discuss these in detail when we encounter them.

4.2.2 Polytopic LDIs

When Ω is a polytope, we will call the LDI a *polytopic LDI* or PLDI. Most of our results require that Ω be described by a list of its vertices, i.e., in the form

$$\mathbf{Co}\left\{\begin{bmatrix} A_1 & B_{u,1} & B_{w,1} \\ C_{z,1} & D_{zu,1} & D_{zw,1} \end{bmatrix}, \ldots, \begin{bmatrix} A_L & B_{u,L} & B_{w,L} \\ C_{z,L} & D_{zu,L} & D_{zw,L} \end{bmatrix}\right\}. \tag{4.8}$$

where the matrices (4.8) are given.

If instead Ω is described by a set of l linear inequalities, then the number of vertices, i.e., L, will generally increase very rapidly (exponentially) with l. Therefore results for PLDIs that require the description (4.8) are of limited interest for problems in which Ω is described in terms of linear inequalities.

4.2.3 Norm-bound LDIs

Another special class of LDIs is the class of norm-bound LDIs (NLDIs), described by

$$\begin{aligned} \dot{x} &= Ax + B_p p + B_u u + B_w w, \quad x(0) = x_0, \\ q &= C_q x + D_{qp} p + D_{qu} u + D_{qw} w, \\ z &= C_z x + D_{zp} p + D_{zu} u + D_{zw} w, \\ p &= \Delta(t) q, \qquad \|\Delta(t)\| \leq 1. \end{aligned} \tag{4.9}$$

where $\Delta : \mathbf{R}_+ \to \mathbf{R}^{n_p \times n_q}$, with $\|\Delta(t)\| \leq 1$ for all t. We will often rewrite the condition $p = \Delta(t)q$, $\|\Delta\| \leq 1$ in the equivalent form $p^T p \leq q^T q$.

For the NLDI (4.9) the set Ω has the form

$$\Omega = \left\{ \tilde{A} + \tilde{B}\Delta\left(I - D_{qp}\Delta\right)^{-1}\tilde{C} \;\middle|\; \|\Delta\| \leq 1 \right\},$$

where

$$\tilde{A} = \begin{bmatrix} A & B_u & B_w \\ C_z & D_{zu} & D_{zw} \end{bmatrix}, \quad \tilde{B} = \begin{bmatrix} B_p \\ D_{zp} \end{bmatrix}, \quad \tilde{C} = \begin{bmatrix} C_q & D_{qu} & D_{qw} \end{bmatrix}. \tag{4.10}$$

Thus, Ω is the image of the (matrix norm) unit ball under a (matrix) linear-fractional mapping. Note that Ω is convex. The set Ω is well defined if and only if $D_{qp}^T D_{qp} < I$, in which case we say that the NLDI is *well-posed*. In the sequel we will consider only well-posed NLDIs.

The equations (4.9) can be interpreted as an LTI system with inputs u, w, output z, and a time-varying feedback matrix $\Delta(t)$ connected between q and p.

Remark: In the sequel, we often drop some terms in (4.9): D_{qw} and D_{zp} will always be zero, and very often D_{qp}, D_{zw}, D_{zu} and D_{qu} will be assumed zero. These assumptions greatly simplify the formulas and treatment. It is not hard to work out the corresponding formulas and results for the more general case.

4.2.4 Diagonal norm-bound LDIs

A useful variation on the NLDI is the diagonal norm-bound LDI (DNLDI), in which we further restrict the matrix Δ to be diagonal. It is given by

$$\begin{aligned} \dot{x} &= Ax + B_p p + B_u u + B_w w, \quad x(0) = x_0, \\ q &= C_q x + D_{qp} p + D_{qu} u + D_{qw} w, \\ z &= C_z x + D_{zp} p + D_{zu} u + D_{zw} w, \\ p_i &= \delta_i(t) q_i, \qquad |\delta_i(t)| \le 1, \quad i = 1, \ldots, n_q. \end{aligned} \tag{4.11}$$

We will often express the condition $p_i = \delta_i(t)q_i$, $|\delta_i(t)| \le 1$ in the simpler form $|p_i| \le |q_i|$. Here we have the componentwise inequalities $|p_i| \le |q_i|$ instead of the single inequality $\|p\| \le \|q\|$ appearing in the NLDI. In the language of control theory, we describe a DNLDI as a linear system with n_q scalar, uncertain, time-varying feedback gains, each of which is bounded by one. The DNLDI is described by the set

$$\Omega = \left\{ \tilde{A} + \tilde{B}\Delta\left(I - D_{qp}\Delta\right)^{-1}\tilde{C} \;\middle|\; \|\Delta\| \le 1, \;\; \Delta \text{ diagonal} \right\}, \tag{4.12}$$

where $\tilde{A}$, $\tilde{B}$ and $\tilde{C}$ are given by (4.10).

The condition of well-posedness for a DNLDI is that $I - D_{qp}\Delta$ should be invertible for all diagonal Δ with $\|\Delta\| \le 1$. This is equivalent to $\det(I - D_{qp}\Delta) > 0$ when Δ takes on its 2^{n_q} extreme values, i.e., $|\Delta_{ii}| = 1$. Although this condition is explicit, the number of inequalities grows exponentially with n_q. In fact, determining whether a DNLDI is well-posed is NP-hard; see the Notes and References.

It turns out that for a DNLDI, $\mathbf{Co}\,\Omega$ is a polytope. In fact,

$$\mathbf{Co}\,\Omega = \mathbf{Co}\left\{ \tilde{A} + \tilde{B}\Delta\left(I - D_{qp}\Delta\right)^{-1}\tilde{C} \;\middle|\; \Delta \text{ diagonal}, \; |\Delta_{ii}| = 1 \right\},$$

i.e., the vertices of $\mathbf{Co}\,\Omega$ are among 2^{n_q} images of the extreme points of Δ under the matrix linear-fractional mapping. Therefore a DNLDI can be described as a PLDI, at least in principle. But there is a very important difference between the two descriptions of the same LDI: the number of vertices required to give the PLDI representation of an LDI increases exponentially with n_q (see the Notes and References of Chapter 5 for more discussion).

Remark: More elaborate variations on the NLDI are sometimes useful. Doyle introduced the idea of "structured feedback," in which the matrix Δ is restricted to have a given block-diagonal form, and in addition, there can be equality constraints among some of the blocks. When we have a single block, we have an NLDI; when it is diagonal (with no equality constraints), we have a DNLDI. Most of the results we present for NLDIs and DNLDIs can be extended to these more general types of LDIs.

4.3 Nonlinear System Analysis via LDIs

Much of our motivation for studying LDIs comes from the fact that we can use them to establish various properties of nonlinear, time-varying systems using a technique known as "global linearization". The idea is implicit in the work on absolute stability originating in the Soviet Union in the 1940s.

Consider the system

$$\dot{x} = f(x, u, w, t), \qquad z = g(x, u, w, t). \tag{4.13}$$

Suppose that for each x, u, w, and t there is a matrix $G(x,u,w,t) \in \Omega$ such that

$$\begin{bmatrix} f(x,u,w,t) \\ g(x,u,w,t) \end{bmatrix} = G(x,u,w,t) \begin{bmatrix} x \\ u \\ w \end{bmatrix} \tag{4.14}$$

where $\Omega \subseteq \mathbf{R}^{(n+n_z)\times(n+n_u+n_w)}$. Then of course every trajectory of the nonlinear system (4.13) is also a trajectory of the LDI defined by Ω. If we can prove that every trajectory of the LDI defined by Ω has some property (e.g., converges to zero), then a fortiori we have proved that every trajectory of the nonlinear system (4.13) has this property.

4.3.1 A derivative condition

Conditions that guarantee the existence of such a G are $f(0,0,0,0) = 0$, $g(0,0,0,0) = 0$, and

$$\begin{bmatrix} \dfrac{\partial f}{\partial x} & \dfrac{\partial f}{\partial u} & \dfrac{\partial f}{\partial w} \\ \dfrac{\partial g}{\partial x} & \dfrac{\partial g}{\partial u} & \dfrac{\partial g}{\partial w} \end{bmatrix} \in \Omega \ \text{ for all } x,\ u,\ w,\ t. \tag{4.15}$$

In fact we can make a stronger statement that links the difference between a pair of trajectories of the nonlinear system and the trajectories of the LDI given by Ω. Suppose we have (4.15) but not necessarily $f(0,0,0,0) = 0$, $g(0,0,0,0) = 0$. Then for any pair of trajectories (x,w,z) and $(\tilde{x},\tilde{w},\tilde{z})$ we have

$$\begin{bmatrix} \dot{x} - \dot{\tilde{x}} \\ z - \tilde{z} \end{bmatrix} \in \mathbf{Co}\,\Omega \begin{bmatrix} x - \tilde{x} \\ u - \tilde{u} \\ w - \tilde{w} \end{bmatrix},$$

i.e., $(x-\tilde{x}, w-\tilde{w}, z-\tilde{z})$ is a trajectory of the LDI given by $\mathbf{Co}\,\Omega$.

These results follow from a simple extension of the mean-value theorem: if $\phi : \mathbf{R}^n \to \mathbf{R}^n$ satisfies

$$\frac{\partial \phi}{\partial x} \in \Omega$$

throughout $\mathbf{R}^n$, then for any x and $\tilde{x}$ we have

$$\phi(x) - \phi(\tilde{x}) \in \mathbf{Co}\,\Omega(x - \tilde{x}). \tag{4.16}$$

To see this, let $c \in \mathbf{R}^n$. By the mean-value theorem we have

$$c^T(\phi(x) - \phi(\tilde{x})) = c^T \frac{\partial \phi}{\partial x}(\zeta)(x - \tilde{x})$$

for some ζ that lies on the line segment between x and $\tilde{x}$. Since by assumption

$$\frac{\partial \phi}{\partial x}(\zeta) \in \Omega,$$

we conclude that

$$c^T(\phi(x) - \phi(\tilde{x})) \le \sup_{A \in \mathbf{Co}\,\Omega} c^T A(x - \tilde{x}).$$

Since c was arbitrary, we see that $\phi(x) - \phi(\tilde{x})$ is in every half-space that contains the convex set $\mathbf{Co}\,\Omega(x - \tilde{x})$, which proves (4.16).

4.3.2 Sector conditions

Sometimes the nonlinear system (4.13) can be expressed in the form

$$\begin{aligned}\dot{x} &= Ax + B_p p + B_u u + B_w w,\\ q &= C_q x + D_{qp} p + D_{qu} u + D_{qw} w,\\ z &= C_z x + D_{zp} p + D_{zu} u + D_{zw} w,\\ p_i &= \phi_i(q_i, t), \quad i = 1, \dots, n_q,\end{aligned} \tag{4.17}$$

where $\phi_i : \mathbf{R} \times \mathbf{R}_+ \to \mathbf{R}$ satisfy the *sector conditions*

$$\alpha_i q^2 \le q\phi_i(q,t) \le \beta_i q^2$$

for all q and $t \ge 0$, where α_i and β_i are given. In words, the system consists of a linear part together with n_q time-varying sector-bounded scalar nonlinearities.

The variables p_i and q_i can be eliminated from (4.17) provided the matrix $I - D_{qp}\Delta$ is nonsingular for all diagonal Δ with $\alpha_i \le \Delta_{ii} \le \beta_i$. In this case, we say the system is well-posed.

Assume now that the system (4.17) is well-posed, and let (4.13) be the equations obtained by eliminating the variables p and q. Define

$$\Omega = \left\{ \tilde{A} + \tilde{B}\Delta\left(I - D_{qp}\Delta\right)^{-1}\tilde{C} \;\middle|\; \Delta \text{ diagonal, } \Delta_{ii} \in [\alpha_i, \beta_i] \right\},$$

where $\tilde{A}$, $\tilde{B}$ and $\tilde{C}$ are given by (4.10). Then, the condition (4.14) holds.

Notes and References

Differential inclusions

The books by Aubin and Cellina [AC84], Kisielewicz [KIS91], and the classic text by Filippov [FIL88] cover the theory of differential inclusions. See also the article by Roxin [ROX65] and references therein. Background material, e.g., set-valued analysis, can be found in the book by Aubin and Frankowska [AF90].

The term *linear differential inclusion* is a bit misleading, since LDIs do not enjoy any particular linearity or superposition properties. We use the term only to point out the interpretation as an uncertain time-varying linear system. Many authors refer to LDIs using some subset of the phrase "uncertain linear time-varying system". A more accurate term, suggested to us and used by A. Filippov and E. Pyatnitskii, is *selector-linear differential inclusion.*

PLDIs come up in several articles on control theory, e.g., [MP89, KP87A, KP87B, BY89].

Integral quadratic constraints

The pointwise quadratic constraint $p(t)^T p(t) \le q(t)^T q(t)$ that occur in NLDIs can be generalized to the integral quadratic constraint $\int_0^t p^T p \, d\tau \le \int_0^t q^T q \, d\tau$; see §8.2 and §8.3.

Global linearization

The idea of replacing a nonlinear system by a time-varying linear system can be found in Liu et al. [LIU68, LSL69]. They call this approach *global linearization.* Of course, approximating the set of trajectories of a nonlinear system via LDIs can be very conservative, i.e., there are many trajectories of the LDI that are not trajectories of the nonlinear system. We will see in Chapter 8 how this conservatism can be reduced in some special cases.

The idea of global linearization is implicit in the early Soviet literature on the absolute stability problem, e.g., Lur'e and Postnikov [LP44, LUR57] and Popov [POP73].

Modeling systems as PLDIs

One of the key issues in robust control theory is how to model or measure plant "uncertainty" or "variation". We propose a simple method that should work well in some cases. At the least, the method is extremely simple and natural.

Suppose we have a real system that is fairly well modeled as a linear system. We collect many sets of input/output measurements, obtained at different times, under different operating conditions, or perhaps from different instances of the system (e.g., different units from a manufacturing run). It is important that we have data sets from enough plants or plant conditions to characterize or at least give some idea of the plant variation that can be expected.

For each data set we develop a linear system model of the plant. To simplify the problem we will assume that the state in this model is accessible, so the different models refer to the same state vector. These models should be fairly close, but of course not exactly the same. We might find for example that the transfer matrices of these models at $s = 0$ differ by about 10%, but at high frequencies they differ considerably more. More importantly, this collection of models contains information about the "structure" of the plant variation.

We propose to model the system as a PLDI with the vertices given by the measured or estimated linear system models. In other words, we model the plant as a time-varying linear system, with system matrices that can jump around among any of the models we estimated. It seems reasonable to conjecture that a controller that works well with this PLDI is likely to work well when connected to the real system. Such controllers can be designed by the methods described in Chapter 7.

Well-posedness of DNLDIs

See [BY89] for a proof that a DNLDI is well-posed if and only if $\det(I - D_{qp}\Delta) > 0$ for every Δ with $|\Delta_{ii}| = 1$; the idea behind this proof can be traced back to Zadeh and Desoer [ZD63, §9.17]; see also [SAE86]. An equivalent condition is in terms of $\mathcal{P}_0$ matrices (a matrix is $\mathcal{P}_0$ if every principal minor is nonnegative; see [FP62, FP66, FP67]): A DNLDI is well-posed if and only if $I + D_{qp}$ is invertible, and $(I + D_{qp})^{-1}(I - D_{qp})$ is a $\mathcal{P}_0$ matrix [GHA90]. Another equivalent condition is that $(I + D_{qp}, I - D_{qp})$ is a $\mathcal{W}_0$ pair (see [WIL70]). The problem of determining whether a general matrix is $\mathcal{P}_0$ is thought to be very difficult, i.e., of high complexity [CPS92, P149].

Standard branch-and-bound techniques can be applied to the problem of determining well-posedness of DNLDIs; see for example [SD86, GS88, SP89, BBB91, BB92]. Of course these methods have no theoretical advantage over simply checking that the 2^{n_q} determinants are positive. But in practice they may be more efficient, especially when used with computationally cheap sufficient conditions (see below).

There are several sufficient conditions for well-posedness of a DNLDI that can be checked in polynomial-time. Fan, Tits, and Doyle [FTD91] show that if the LMI

$$D_{qp}^T P D_{qp} < P, \qquad P \text{ diagonal}, \qquad P > 0, \tag{4.18}$$

is feasible, then the DNLDI is well-posed. Condition (4.18) is equivalent to the existence of a diagonal (scaling) matrix R such that $\|R D_{qp} R^{-1}\| < 1$. In [FTD91], the authors also give a sufficient condition for well-posedness of more general "structured" LDIs, e.g., when equality constraints are imposed on the matrix Δ in (4.11). These conditions can also be obtained using the $\mathcal{S}$-procedure; see [FER93].

Representing DNLDIs as PLDIs

For a well-posed DNLDI, $\mathbf{Co}\,\Omega$ is a polytope, which means that the DNLDI (4.12) can also be represented as a PLDI. More specifically, we have

$$\mathbf{Co}\,\Omega = \mathbf{Co}\left\{ \tilde{A} + \tilde{B}\Delta\,(I - D_{qp}\Delta)^{-1}\,\tilde{C} \;\middle|\; \Delta \text{ diagonal}, \, |\Delta_{ii}| = 1 \right\},$$

where $\tilde{A}$, $\tilde{B}$ and $\tilde{C}$ are given by (4.10). Note that the set on the right-hand side is a polytope with at most 2^{n_q} vertices. We now prove this result.

Define

$$\mathcal{K} = \left\{ \tilde{A} + \tilde{B}\Delta \left(I - D_{qp}\Delta\right)^{-1} \tilde{C} \;\middle|\; \Delta \text{ diagonal}, |\Delta_{ii}| = 1 \right\},$$

which is a set containing at most 2^{n_q} points. By definition

$$\mathbf{Co}\,\Omega = \mathbf{Co}\left\{ \tilde{A} + \tilde{B}\Delta \left(I - D_{qp}\Delta\right)^{-1} \tilde{C} \;\middle|\; \Delta \text{ diagonal}, |\Delta_{ii}| \le 1 \right\}.$$

Clearly we have $\mathbf{Co}\,\Omega \supseteq \mathbf{Co}\,\mathcal{K}$. We will now show that every extreme point of $\mathbf{Co}\,\Omega$ is contained in $\mathbf{Co}\,\mathcal{K}$, which will imply that $\mathbf{Co}\,\mathcal{K} \supseteq \mathbf{Co}\,\Omega$, completing the proof.

Let W_{ext} be an extreme point of $\mathbf{Co}\,\Omega$. Then there exists a linear function ψ such that W_{ext} is the unique maximizer of ψ over $\mathbf{Co}\,\Omega$. In fact, $W_{\text{ext}} \in \Omega$, since otherwise, the half-space $\{W \mid \psi(W) < \psi(W_{\text{ext}}) - \epsilon\}$, where $\epsilon > 0$ is sufficiently small, contains Ω but not W_{ext}, which contradicts $W_{\text{ext}} \in \mathbf{Co}\,\Omega$.

Now, for $W \in \Omega$, $\psi(W)$ is a bounded linear-fractional function of the first diagonal entry Δ_{11} of Δ, for fixed $\Delta_{22}, \ldots, \Delta_{n_q n_q}$, i.e., it equals $(a + b\Delta_{11})/(c + d\Delta_{11})$ for some a, b, c and d. Moreover, the denominator $c + d\Delta_{11}$ is nonzero for $\Delta_{11} \in [-1, 1]$, since the DNLDI is well-posed. Therefore, for every fixed $\Delta_{22}, \ldots, \Delta_{n_q n_q}$, the function ψ achieves a maximum at an extreme value of Δ_{11}, i.e., $\Delta_{11} = \pm 1$. Extending this argument to $\Delta_{22}, \ldots, \Delta_{n_q n_q}$ leads us to the following conclusion: $W_{\text{ext}} = \tilde{A} + \tilde{B}S\left(I - D_{qp}S\right)^{-1}\tilde{C}$, where S satisfies $S_{ii} = \pm 1$. In other words, $W_{\text{ext}} \in \mathbf{Co}\,\mathcal{K}$, which concludes our proof.

We have already noted that the description of a DNLDI as a PLDI will be much larger than its description as a DNLDI; see the Notes and References of Chapter 5, and also [BY89] and [PS82, KP87A, KAM83].

Approximating PLDIs by NLDIs

It is often useful to conservatively approximate a PLDI as an NLDI. One reason, among many, is the potentially much smaller size of the description as an NLDI compared to the description of the PLDI. We consider the PLDI

$$\dot{x} = A(t)x, \qquad A(t) \in \Omega_{\text{PLDI}} \stackrel{\Delta}{=} \mathbf{Co}\left\{A_1, \ldots, A_L\right\}, \tag{4.19}$$

with $x(t) \in \mathbf{R}^n$, and the NLDI

$$\dot{x} = (A + B_p\Delta(t)C_q)x, \qquad \|\Delta(t)\| \le 1, \tag{4.20}$$

with associated set $\Omega_{\text{NLDI}} \stackrel{\Delta}{=} \{A + B_p\Delta C_q \mid \|\Delta\| \le 1\}$. Our goal is to find A, B_p, and C_q such that $\Omega_{\text{PLDI}} \subseteq \Omega_{\text{NLDI}}$ with the set Ω_{NLDI} as small as possible in some appropriate sense. This will give an efficient outer approximation of the PLDI (4.19) by the NLDI (4.20). Since $\Omega_{\text{PLDI}} \subseteq \Omega_{\text{NLDI}}$ implies that every trajectory of the PLDI is a trajectory of the NLDI, every result we establish for all trajectories of the NLDI holds for all trajectories of the PLDI. For L very large, it is much easier to work with the NLDI than the PLDI.

For simplicity, we will only consider the case with $B_p \in \mathbf{R}^{n\times n}$ and invertible. Evidently there is substantial redundancy in our representation of the NLDI (4.20): We can replace B_p by αB_p and C_q by C_q/α where $\alpha \ne 0$. (We can also replace B_p and C_q with B_pU and VC_q where U and V are any orthogonal matrices, without affecting the set Ω_{NLDI}, but we will not use this fact.)

We have $\Omega_{\text{PLDI}} \subseteq \Omega_{\text{NLDI}}$ if for every ξ and $k = 1, \ldots, L$, there exists π such that

$$B_p\pi = (A_k - A)\xi, \qquad \pi^T\pi \le \xi^T C_q^T C_q \xi.$$

This yields the equivalent condition

$$(A_k - A)^T B_p^{-T} B_p^{-1} (A_k - A) \le C_q^T C_q, \quad k = 1, \ldots, L,$$

which, in turn, is equivalent to

$$B_p B_p^T > 0, \qquad \begin{bmatrix} C_q^T C_q & (A_k - A)^T \\ A_k - A & B_p B_p^T \end{bmatrix} \ge 0, \quad k = 1, \ldots, L.$$

Introducing new variables $V = C_q^T C_q$ and $W = B_p B_p^T$, we get the equivalent LMI in A, V and W

$$W > 0, \qquad \begin{bmatrix} V & (A_k - A)^T \\ A_k - A & W \end{bmatrix} \ge 0, \quad k = 1, \ldots, L. \tag{4.21}$$

Thus we have $\Omega_{\text{PLDI}} \subseteq \Omega_{\text{NLDI}}$ if condition (4.21) holds.

There are several ways to minimize the size of $\Omega_{\text{NLDI}} \supseteq \Omega_{\text{PLDI}}$. The most obvious is to minimize $\mathbf{Tr}\, V + \mathbf{Tr}\, W$ subject to (4.21), which is an EVP in A, V and W. This objective is clearly related to several measures of the size of Ω_{NLDI}, but we do not know of any simple interpretation.

We can formulate the problem of minimizing the *diameter* of Ω_{NLDI} as an EVP. We define the diameter of a set $\Omega \subseteq \mathbf{R}^{n \times n}$ as $\max \{\|F - G\| \mid F,\ G \in \Omega\}$. The diameter of Ω_{NLDI} is equal to $2\sqrt{\lambda_{\max}(V)\lambda_{\max}(W)}$. We can exploit the scaling redundancy in C_q to assume without loss of generality that $\lambda_{\max}(V) \le 1$, and then minimize $\lambda_{\max}(W)$ subject to (4.21) and $V \le I$, which is an EVP. It can be shown that the resulting optimal V^{opt} satisfies $\lambda_{\max}(V^{\text{opt}}) = 1$, so that we have in fact minimized the diameter of Ω_{NLDI} subject to $\Omega_{\text{PLDI}} \subseteq \Omega_{\text{NLDI}}$.

Chapter 5

Analysis of LDIs: State Properties

In this chapter we consider properties of the state x of the LDI

$$\dot{x} = A(t)x, \qquad A(t) \in \Omega, \tag{5.1}$$

where $\Omega \subseteq \mathbf{R}^{n\times n}$ has one of four forms:

- LTI SYSTEMS: LTI systems are described by $\dot{x} = Ax$.
- POLYTOPIC LDIS: PLDIs are described by $\dot{x} = A(t)x$, $A(t) \in \mathbf{Co}\{A_1, \dots, A_L\}$.
- NORM-BOUND LDIS: NLDIs are described by

$$\dot{x} = Ax + B_p p, \qquad q = C_q x + D_{qp} p, \qquad p = \Delta(t) q, \qquad \|\Delta(t)\| \le 1,$$

which we will rewrite as

$$\dot{x} = Ax + B_p p, \qquad p^T p \le (C_q x + D_{qp} p)^T (C_q x + D_{qp} p). \tag{5.2}$$

We assume well-posedness, i.e., $\|D_{qp}\| < 1$.

- DIAGONAL NORM-BOUND LDIS: DNLDIs are described by

$$\begin{array}{ll} \dot{x} = Ax + B_p p, & q = C_q x + D_{qp} p, \\ p_i = \delta_i(t) q_i, & |\delta_i(t)| \le 1, \quad i = 1, \dots, n_q. \end{array} \tag{5.3}$$

which can be rewritten as

$$\dot{x} = Ax + B_p p, \qquad q = C_q x + D_{qp} p, \qquad |p_i| \le |q_i|, \quad i = 1, \dots, n_q.$$

Again, we assume well-posedness.

5.1 Quadratic Stability

We first study *stability* of the LDI (5.1), that is, we ask whether all trajectories of system (5.1) converge to zero as $t \to \infty$. A sufficient condition for this is the existence of a quadratic function $V(\xi) = \xi^T P \xi$, $P > 0$ that decreases along every nonzero trajectory of (5.1). If there exists such a P, we say the LDI (5.1) is *quadratically stable* and we call V a *quadratic Lyapunov function*.

Since

$$\frac{d}{dt}V(x) = x^T \left(A(t)^T P + PA(t)\right) x,$$

a necessary and sufficient condition for quadratic stability of system (5.1) is

$$P > 0, \qquad A^T P + PA < 0 \quad \text{for all } A \in \Omega. \tag{5.4}$$

Multiplying the second inequality in (5.4) on the left and right by P^{-1}, and defining a new variable $Q = P^{-1}$, we may rewrite (5.4) as

$$Q > 0, \qquad QA^T + AQ < 0 \quad \text{for all } A \in \Omega. \tag{5.5}$$

This dual inequality is an equivalent condition for quadratic stability. We now show that conditions for quadratic stability for LTI systems, PLDIs, and NLDIs can be expressed in terms of LMIs.

• LTI SYSTEMS: Condition (5.4) becomes

$$P > 0, \qquad A^T P + PA < 0. \tag{5.6}$$

Therefore, checking quadratic stability for an LTI system is an LMIP in the variable P. This is precisely the (necessary and sufficient) Lyapunov stability criterion for LTI systems. (In other words, a linear system is stable if and only if it is quadratically stable.) Alternatively, using (5.5), stability of LTI systems is equivalent to the existence of Q satisfying the LMI

$$Q > 0, \qquad AQ + QA^T < 0. \tag{5.7}$$

Of course, each of these (very special) LMIPs can be solved analytically by solving a Lyapunov equation (see §1.2 and §2.7).

• POLYTOPIC LDIs: Condition (5.4) is equivalent to

$$P > 0, \qquad A_i^T P + PA_i < 0, \quad i = 1, \ldots, L. \tag{5.8}$$

Thus, determining quadratic stability for PLDIs is an LMIP in the variable P. The dual condition (5.5) is equivalent to the LMI in the variable Q

$$Q > 0, \qquad QA_i^T + A_i Q < 0, \quad i = 1, \ldots, L. \tag{5.9}$$

• NORM-BOUND LDIs: Condition (5.4) is equivalent to $P > 0$ and

$$\begin{bmatrix} \xi \\ \pi \end{bmatrix}^T \begin{bmatrix} A^T P + PA & PB_p \\ B_p^T P & 0 \end{bmatrix} \begin{bmatrix} \xi \\ \pi \end{bmatrix} < 0 \tag{5.10}$$

for all nonzero ξ satisfying

$$\begin{bmatrix} \xi \\ \pi \end{bmatrix}^T \begin{bmatrix} -C_q^T C_q & -C_q^T D_{qp} \\ -D_{qp}^T C_q & I - D_{qp}^T D_{qp} \end{bmatrix} \begin{bmatrix} \xi \\ \pi \end{bmatrix} \leq 0. \tag{5.11}$$

In order to apply the $\mathcal{S}$-procedure we show that the set

$$\mathcal{A} = \{(\xi, \pi) \mid \xi \neq 0, \quad (5.11)\}$$

equals the set

$$\mathcal{B} = \{(\xi,\pi) \mid (\xi,\pi) \neq 0, \quad (5.11)\}.$$

It suffices to show that $\{(\xi,\pi) \mid \xi = 0, \quad \pi \neq 0, \quad (5.11)\} = \emptyset$. But this is immediate: If $\pi \neq 0$, then condition (5.11) cannot hold without having $\xi \neq 0$, since $I - D_{qp}^T D_{qp} > 0$. Therefore, the condition that $dV(x)/dt < 0$ for all nonzero trajectories is equivalent to (5.10) being satisfied for any nonzero (ξ,π) satisfying (5.11). (This argument recurs throughout this chapter and will not be repeated.) Using the $\mathcal{S}$-procedure, we find that quadratic stability of (5.2) is equivalent to the existence of P and λ satisfying

$$P > 0, \qquad \lambda \geq 0,$$
$$\begin{bmatrix} A^T P + PA + \lambda C_q^T C_q & PB_p + \lambda C_q^T D_{qp} \\ (PB_p + \lambda C_q^T D_{qp})^T & -\lambda(I - D_{qp}^T D_{qp}) \end{bmatrix} < 0. \tag{5.12}$$

Thus, determining quadratic stability of an NLDI is an LMIP. Since LMI (5.12) implies that $\lambda > 0$, we may, by defining $\tilde{P} = P/\lambda$, obtain an equivalent condition

$$\tilde{P} > 0, \qquad \begin{bmatrix} A^T \tilde{P} + \tilde{P} A + C_q^T C_q & \tilde{P} B_p + C_q^T D_{qp} \\ (\tilde{P} B_p + C_q^T D_{qp})^T & -(I - D_{qp}^T D_{qp}) \end{bmatrix} < 0, \tag{5.13}$$

an LMI in the variable $\tilde{P}$. Thus, quadratic stability of the NLDI has the frequency-domain interpretation that the $\mathbf{H}_\infty$ norm of the LTI system

$$\dot{x} = Ax + B_p p, \qquad q = C_q x + D_{qp} p$$

is less than one.

With the new variables $Q = P^{-1}$, $\mu = 1/\lambda$, quadratic stability of NLDIs is also equivalent to the existence of μ and Q satisfying the LMI

$$\mu \geq 0, \qquad Q > 0,$$
$$\begin{bmatrix} AQ + QA^T + \mu B_p B_p^T & \mu B_p D_{qp}^T + QC_q^T \\ (\mu B_p D_{qp}^T + QC_q^T)^T & -\mu(I - D_{qp} D_{qp}^T) \end{bmatrix} < 0. \tag{5.14}$$

Remark: Note that our assumption of well-posedness is in fact incorporated in the LMI (5.12) since it implies $I - D_{qp} D_{qp}^T > 0$. Therefore the NLDI (5.2) is well-posed and quadratically stable if and only if the LMIP (5.12) has a solution.

In the remainder of this chapter we will assume that D_{qp} in (5.2) is zero; the reader should bear in mind that all the following results hold for nonzero D_{qp} as well.

• DIAGONAL NORM-BOUND LDIS: For the DNLDI (5.3), we can obtain only a sufficient condition for quadratic stability. The condition $dV(x)/dt < 0$ for all nonzero trajectories is equivalent to

$$\xi^T (A^T P + PA)\xi + 2\xi^T P B_p \pi < 0$$

for all nonzero (ξ,π) satisfying

$$\pi_i^2 \leq (C_{q,i}\xi + D_{qp,i}\pi)^T (C_{q,i}\xi + D_{qp,i}\pi), \qquad i = 1,\ldots,L,$$

where we have used $C_{q,i}$ and $D_{qp,i}$ to denote the ith rows of C_q and D_{qp} respectively. Using the $\mathcal{S}$-procedure, we see that this condition is implied by the existence of nonnegative $\lambda_1, \ldots, \lambda_{n_q}$ such that

$$\xi^T(A^TP + PA)\xi + 2\xi^TPB_p\pi + \sum_{i=1}^{L} \lambda_i \left((C_{q,i}\xi + D_{qp,i}\pi)^T (C_{q,i}\xi + D_{qp,i}\pi) - \pi_i^2 \right) < 0$$

for all nonzero (ξ, π). With $\Lambda = \mathbf{diag}(\lambda, \ldots, \lambda_{n_q})$, this is equivalent to

$$\begin{bmatrix} A^TP + PA + C_q^T\Lambda C_q & PB_p + C_q^T\Lambda D_{qp} \\ B_p^TP + D_{qp}^T\Lambda C_q & D_{qp}^T\Lambda D_{qp} - \Lambda \end{bmatrix} < 0. \tag{5.15}$$

Therefore, if there exist $P > 0$ and diagonal $\Lambda \geq 0$ satisfying (5.15), then the DNLDI (5.3) is quadratically stable. Checking this sufficient condition for quadratic stability is an LMIP. Note that from (5.15), we must have $\Lambda > 0$.

Equivalently, quadratic stability is implied by the existence of $Q > 0$, $M = \mathbf{diag}(\mu_1, \ldots, \mu_{n_q}) > 0$ satisfying the LMI

$$\begin{bmatrix} AQ + QA^T + B_pMB_p^T & QC_q^T + B_pMD_{qp}^T \\ C_qQ + D_{qp}MB_p^T & -M + D_{qp}MD_{qp}^T \end{bmatrix} < 0, \tag{5.16}$$

another LMIP.

Remark: Quadratic stability of a DNLDI via the $\mathcal{S}$-procedure has a simple frequency-domain interpretation. Denoting by H the transfer matrix $H(s) = D_{qp} + C_q(sI - A)^{-1}B_p$, and assuming (A, B_p, C_q) is minimal, quadratic stability is equivalent to the fact that, for some diagonal, positive-definite matrix Λ, $\|\Lambda^{1/2}H\Lambda^{-1/2}\|_\infty < 1$. $\Lambda^{-1/2}$ can then be interpreted as a *scaling*. (See the Notes and References for more details on the connection between the $\mathcal{S}$-procedure and scaling.)

Remark: Here too our assumption of well-posedness is in fact incorporated in the LMI (5.15) or (5.16). The bottom right block is precisely the sufficient condition for well-posedness mentioned on page 57. Therefore the DNLDI (5.2) is well-posed and quadratically stable if the LMI (5.15) or (5.16) is feasible.

Note the similarity between the corresponding LMIs for the NLDI and the DNLDI; the only difference is that the diagonal matrix appearing in the LMI associated with the DNLDI is fixed as the scaled identity matrix in the LMI associated with the NLDI. In general it is straightforward to derive the corresponding (sufficient) condition for DNLDIs from the result for NLDIs. In the remainder of the book, we will often restrict our attention to NLDIs, and pay only occasional attention to DNLDIs; we leave to the reader the task of generalizing all the results for NLDIs to DNLDIs.

Remark: An LTI system is stable if and only if it is quadratically stable. For more general LDIs, however, stability does not imply quadratic stability (see the Notes and References at the end of this chapter). It is true that an LDI is stable if and only if there exists a *convex* Lyapunov function that proves it; see the Notes and References.

5.1.1 Coordinate transformations

We can give another interpretation of quadratic stability in terms of state-space coordinate transformations. Consider the change of coordinates $x = T\bar{x}$, where $\det T \neq 0$. In the new coordinate system, (5.1) is described by the matrix $\bar{A}(t) = T^{-1}A(t)T$. We ask the question: Does there exist T such that in the new coordinate system, all (nonzero) trajectories of the LDI are always decreasing in norm, i.e., $d\|\bar{x}\|/dt < 0$? It is easy to show that this is true if and only if there exists a quadratic Lyapunov function $V(\xi) = \xi^T P\xi$ for the LDI, in which case we can take any T with $TT^T = P^{-1}$, for example, $T = P^{-1/2}$ (these T's are all related by right orthogonal matrix multiplication). With this interpretation, it is natural to seek a coordinate transformation matrix T that makes all nonzero trajectories decreasing in norm, and has the smallest possible condition number $\kappa(T)$. This turns out to be an EVP for LTI systems, polytopic and norm-bound LDIs.

To see this, let $P = T^{-T}T^{-1}$, so that $\kappa(T)^2 = \kappa(P)$. Minimizing the condition number of T subject to the requirement that $d\|\bar{x}\|/dt < 0$ for all nonzero trajectories is then equivalent to minimizing the condition number of P subject to $d(x^T Px)/dt < 0$. (Any T with $TT^T = P_{\rm opt}^{-1}$ where $P_{\rm opt}$ is an optimal solution, is optimal for the original problem.) For each of our systems (LTI systems, PLDIs and NLDIs), the change of coordinates with smallest condition number is obtained by solving the following EVPs:

- LTI SYSTEMS: The EVP in the variables P and η is

$$\begin{array}{ll} \text{minimize} & \eta \\ \text{subject to} & (5.6), \qquad I \leq P \leq \eta I \end{array}$$

(In this formulation, we have taken advantage of the homogeneity of the LMI (5.6) in the variable P.)

- POLYTOPIC LDIS: The EVP in the variables P and η is

$$\begin{array}{ll} \text{minimize} & \eta \\ \text{subject to} & (5.8), \qquad I \leq P \leq \eta I \end{array}$$

- NORM-BOUND LDIS: The EVP in the variables P, η and λ is

$$\begin{array}{ll} \text{minimize} & \eta \\ \text{subject to} & (5.12), \qquad I \leq P \leq \eta I \end{array}$$

5.1.2 Quadratic stability margins

Quadratic stability margins give a measure of how much the set Ω can be expanded about some center with the LDI remaining quadratically stable.

- POLYTOPIC LDIS: For the PLDI

$$\dot{x} = (A_0 + A(t))x, \qquad A(t) \in \alpha\ \mathbf{Co}\{A_1, \ldots, A_L\},$$

we define the *quadratic stability margin* as the largest nonnegative α for which it is quadratically stable. This quantity is computed by solving the following GEVP in P

and α:

$$\begin{array}{ll} \text{maximize} & \alpha \\ \text{subject to} & P>0, \qquad \alpha \geq 0, \\ & A_0^T P + PA_0 + \alpha\left(A_i^T P + PA_i\right) < 0, \quad i=1,\ldots,L \end{array}$$

• NORM-BOUND LDIs: The quadratic stability margin of the system

$$\dot{x} = Ax + B_p p, \qquad p^T p \leq \alpha^2 x^T C_q^T C_q x,$$

is defined as the largest $\alpha \geq 0$ for which the system is quadratically stable, and is computed by solving the GEVP in P, λ and $\beta = \alpha^2$:

$$\begin{array}{ll} \text{maximize} & \beta \\ \text{subject to} & P>0, \qquad \beta \geq 0, \qquad \begin{bmatrix} A^T P + PA + \beta\lambda C_q^T C_q & PB_p \\ B_p^T P & -\lambda I \end{bmatrix} < 0 \end{array}$$

Defining $\tilde{P} = P/\lambda$, we get an equivalent EVP in $\tilde{P}$ and β:

$$\begin{array}{ll} \text{maximize} & \beta \\ \text{subject to} & \tilde{P}>0, \qquad \beta \geq 0, \qquad \begin{bmatrix} A^T \tilde{P} + \tilde{P}A + \beta C_q^T C_q & \tilde{P}B_p \\ B_p^T \tilde{P} & -I \end{bmatrix} < 0 \end{array}$$

Remark: The quadratic stability margin obtained by solving this EVP is just $1/\|C_q(sI-A)^{-1}B_p\|_\infty$.

5.1.3 Decay rate

The *decay rate* (or largest Lyapunov exponent) of the LDI (5.1) is defined to be the largest α such that

$$\lim_{t\to\infty} e^{\alpha t}\|x(t)\| = 0$$

holds for all trajectories x. Equivalently, the decay rate is the supremum of

$$\liminf_{t\to\infty} \frac{-\log\|x(t)\|}{t}$$

over all nonzero trajectories. (Stability corresponds to positive decay rate.)

We can use the quadratic Lyapunov function $V(\xi) = \xi^T P \xi$ to establish a lower bound on the decay rate of the LDI (5.1). If $dV(x)/dt \leq -2\alpha V(x)$ for all trajectories, then $V(x(t)) \leq V(x(0))e^{-2\alpha t}$, so that $\|x(t)\| \leq e^{-\alpha t}\kappa(P)^{1/2}\|x(0)\|$ for all trajectories, and therefore the decay rate of the LDI (5.1) is at least α.

• LTI SYSTEMS: The condition that $dV(x)/dt \leq -2\alpha V(x)$ for all trajectories is equivalent to

$$A^T P + PA + 2\alpha P \leq 0. \tag{5.17}$$

Therefore, the largest lower bound on the decay rate that we can find using a quadratic Lyapunov function can be found by solving the following GEVP in P and α:

$$\begin{array}{ll} \text{maximize} & \alpha \\ \text{subject to} & P>0, \qquad (5.17) \end{array} \tag{5.18}$$

This lower bound is sharp, i.e., the optimal value of the GEVP (5.18) is the decay rate of the LTI system (which is the *stability degree* of A, i.e., negative of the maximum real part of the eigenvalues of A).

As an alternate condition, there exists a quadratic Lyapunov function proving that the decay rate is at least α if and only if there exists Q satisfying the LMI

$$Q > 0, \qquad AQ + QA^T + 2\alpha Q \leq 0. \tag{5.19}$$

• POLYTOPIC LDIs: The condition that $dV(x)/dt \leq -2\alpha V(x)$ for all trajectories is equivalent to the LMI

$$A_i^T P + PA_i + 2\alpha P \leq 0, \quad i = 1, \ldots, L. \tag{5.20}$$

Therefore, the largest lower bound on the decay rate provable via quadratic Lyapunov functions is obtained by maximizing α subject to (5.20) and $P > 0$. This is a GEVP in P and α.

As an alternate condition, there exists a quadratic Lyapunov function proving that the decay rate is at least α if and only if there exists Q satisfying the LMI

$$Q > 0, \qquad A_iQ + QA_i^T + 2\alpha Q \leq 0, \quad i = 1, \ldots, L. \tag{5.21}$$

• NORM-BOUND LDIs: Applying the $\mathcal{S}$-procedure, the condition that $dV(x)/dt \leq -2\alpha V(x)$ for all trajectories is equivalent to the existence of $\lambda \geq 0$ such that

$$\begin{bmatrix} A^T P + PA + \lambda C_q^T C_q + 2\alpha P & PB_p \\ B_p^T P & -\lambda I \end{bmatrix} \leq 0. \tag{5.22}$$

Therefore, we obtain the largest lower bound on the decay rate of (5.2) by maximizing α over the variables α, P and λ, subject to $P > 0$, $\lambda \geq 0$ and (5.22), a GEVP.

The lower bound has a simple frequency-domain interpretation: Define the *α-shifted $\mathbf{H}_\infty$ norm* of the system by

$$\|H\|_{\infty,\alpha} \stackrel{\Delta}{=} \sup\left\{ \|H(s)\| \mid \mathbf{Re}\, s > -\alpha \right\}.$$

Then the optimal value of the GEVP is equal to the largest α such that $\|H\|_{\infty,\alpha} < 1$. This can be seen by rewriting (5.22) as

$$\begin{bmatrix} (A + \alpha I)^T P + P(A + \alpha I) + \lambda C_q^T C_q & PB_p \\ B_p^T P & -\lambda I \end{bmatrix} \leq 0,$$

and noting that the $\mathbf{H}_\infty$ norm of the system $(A + \alpha I, B_p, C_q)$ equals $\|H\|_{\infty,\alpha}$.

An alternate necessary and sufficient condition for the existence of a quadratic Lyapunov function proving that the decay rate of (5.2) is at least α is that there exists Q and μ satisfying the LMI

$$\begin{gathered} Q > 0, \qquad \mu \geq 0, \\ \begin{bmatrix} AQ + QA^T + \mu B_p B_p^T + 2\alpha Q & QC_q^T \\ C_q Q & -\mu I \end{bmatrix} \leq 0. \end{gathered} \tag{5.23}$$

5.2 Invariant Ellipsoids

Quadratic stability can also be interpreted in terms of *invariant ellipsoids*. For $Q > 0$, let $\mathcal{E}$ denote the ellipsoid centered at the origin

$$\mathcal{E} = \left\{\xi \in \mathbf{R}^n \mid \xi^T Q^{-1}\xi \leq 1\right\}.$$

The ellipsoid $\mathcal{E}$ is said to be invariant for the LDI (5.1) if for every trajectory x of the LDI, $x(t_0) \in \mathcal{E}$ implies $x(t) \in \mathcal{E}$ for all $t \geq t_0$. It is easily shown that this is the case if and only if Q satisfies

$$QA^T + AQ \leq 0, \quad \text{for all } A \in \Omega,$$

or equivalently,

$$A^T P + PA \leq 0, \quad \text{for all } A \in \Omega, \tag{5.24}$$

where $P = Q^{-1}$. Thus, for LTI systems, PLDIs and NLDIs, invariance of $\mathcal{E}$ is characterized by LMIs in Q or P, its inverse.

Remark: Condition (5.24) is just the nonstrict version of condition (5.4).

- LTI SYSTEMS: The corresponding LMI in P is

$$P > 0, \qquad A^T P + PA \leq 0, \tag{5.25}$$

and the LMI in Q is

$$Q > 0, \qquad AQ + PA^T \leq 0, \tag{5.26}$$

- POLYTOPIC LDIS: The corresponding LMI in P is

$$P > 0, \qquad A_i^T P + PA_i \leq 0, \quad i = 1, \dots, L, \tag{5.27}$$

and the LMI in Q is

$$Q > 0, \qquad QA_i^T + A_i Q \leq 0, \quad i = 1, \dots, L. \tag{5.28}$$

- NORM-BOUND LDIS: Applying the $\mathcal{S}$-procedure, invariance of the ellipsoid $\mathcal{E}$ is equivalent to the existence of λ such that

$$P > 0, \qquad \lambda \geq 0,$$

$$\begin{bmatrix} A^T P + PA + \lambda C_q^T C_q & PB_p + \lambda C_q^T D_{qp} \\ (PB_p + \lambda C_q^T D_{qp})^T & -\lambda(I - D_{qp}^T D_{qp}) \end{bmatrix} \leq 0. \tag{5.29}$$

Invariance of the ellipsoid $\mathcal{E}$ is equivalent to the existence of μ such that

$$\mu \geq 0, \qquad Q > 0, \qquad \text{and}$$

$$\begin{bmatrix} AQ + QA^T + \mu B_p B_p^T & \mu B_p D_{qp}^T + QC_q^T \\ (\mu B_p D_{qp}^T + QC_q^T)^T & -\mu(I - D_{qp} D_{qp}^T) \end{bmatrix} \leq 0. \tag{5.30}$$

In summary, the condition that $\mathcal{E}$ be an invariant ellipsoid can be expressed in each case as an LMI in Q or its inverse P.

5.2.1 Smallest invariant ellipsoid containing a polytope

Consider a polytope described by its vertices, $\mathcal{P} = \mathbf{Co}\{v_1, \ldots, v_p\}$. The ellipsoid $\mathcal{E}$ contains the polytope $\mathcal{P}$ if and only if

$$v_j^T Q^{-1} v_j \le 1, \quad j = 1, \ldots, p.$$

This condition may be expressed as an LMI in Q

$$\begin{bmatrix} 1 & v_j^T \\ v_j & Q \end{bmatrix} \ge 0, \quad j = 1, \ldots, p, \tag{5.31}$$

or as an LMI in $P = Q^{-1}$ as

$$v_j^T P v_j \le 1, \quad j = 1, \ldots, p. \tag{5.32}$$

Minimum volume

The volume of $\mathcal{E}$ is, up to a constant that depends on n, $\sqrt{\det Q}$. We can minimize this volume by solving appropriate CPs.

- LTI SYSTEMS: The CP in P is

$$\begin{array}{ll} \text{minimize} & \log\det P^{-1} \\ \text{subject to} & (5.25), \quad (5.32) \end{array}$$

- POLYTOPIC LDIS: The CP in P is

$$\begin{array}{ll} \text{minimize} & \log\det P^{-1} \\ \text{subject to} & (5.27), \quad (5.32) \end{array} \tag{5.33}$$

- NORM-BOUND LDIS: The CP in P and λ is

$$\begin{array}{ll} \text{minimize} & \log\det P^{-1} \\ \text{subject to} & (5.29), \quad (5.32) \end{array}$$

Minimum diameter

The diameter of the ellipsoid $\mathcal{E}$ is $2\sqrt{\lambda_{\max}(Q)}$. We can minimize this quantity by solving appropriate EVPs.

- LTI SYSTEMS: The EVP in the variables Q and λ is

$$\begin{array}{ll} \text{minimize} & \lambda \\ \text{subject to} & (5.26), \quad (5.31), \quad Q \le \lambda I \end{array}$$

- POLYTOPIC LDIS: The EVP in Q and λ is

$$\begin{array}{ll} \text{minimize} & \lambda \\ \text{subject to} & (5.28), \quad (5.31), \quad Q \le \lambda I \end{array} \tag{5.34}$$

- NORM-BOUND LDIS: The EVP in Q, λ and μ is

$$\begin{array}{ll} \text{minimize} & \lambda \\ \text{subject to} & (5.30), \quad (5.31), \quad Q \le \lambda I \end{array}$$

Remark: These results have the following use. The polytope $\mathcal{P}$ represents our knowledge of the state at time t_0, i.e., $x(t_0) \in \mathcal{P}$ (this knowledge may reflect measurements or prior assumptions). An invariant ellipsoid $\mathcal{E}$ containing $\mathcal{P}$ then gives a bound on the state for $t \geq t_0$, i.e., we can guarantee that $x(t) \in \mathcal{E}$ for all $t \geq t_0$.

5.2.2 Largest invariant ellipsoid contained in a polytope

We now consider a polytope described by linear inequalities:

$$\mathcal{P} = \left\{ \xi \in \mathbf{R}^n \;\middle|\; a_k^T \xi \leq 1, \quad k = 1, \ldots, q \right\}.$$

The ellipsoid $\mathcal{E}$ is contained in the polytope $\mathcal{P}$ if and only if

$$\max \left\{ a_k^T \xi \mid \xi \in \mathcal{E} \right\} \leq 1, \qquad k = 1, \ldots, q.$$

This is equivalent to

$$a_k^T Q a_k \leq 1, \quad k = 1, \ldots, q, \tag{5.35}$$

which is a set of linear inequalities in Q. The maximum volume of invariant ellipsoids contained in $\mathcal{P}$ can be found by solving CPs:

- LTI SYSTEMS: For LTI systems, the CP in the variable Q is

$$\begin{array}{ll} \text{minimize} & \log\det Q^{-1} \\ \text{subject to} & (5.26), \qquad (5.35) \end{array}$$

(Since the volume of $\mathcal{E}$ is, up to a constant, $\sqrt{\det Q}$, minimizing $\log\det Q^{-1}$ will maximize the volume of $\mathcal{E}$).

- POLYTOPIC LDIS: For PLDIs, the CP in the variable Q is

$$\begin{array}{ll} \text{minimize} & \log\det Q^{-1} \\ \text{subject to} & (5.28), \qquad (5.35) \end{array}$$

- NORM-BOUND LDIS: For NLDIs, the CP in the variables Q and μ is

$$\begin{array}{ll} \text{minimize} & \log\det Q^{-1} \\ \text{subject to} & (5.30), \qquad (5.35) \end{array}$$

We also can find the maximum minor diameter (that is, the length of the minor axis) of invariant ellipsoids contained in $\mathcal{P}$ by solving EVPs:

- LTI SYSTEMS: For LTI systems, the EVP in the variables Q and λ is

$$\begin{array}{ll} \text{maximize} & \lambda \\ \text{subject to} & (5.26), \qquad (5.35), \qquad \lambda I \leq Q \end{array}$$

- POLYTOPIC LDIS: For PLDIs, the EVP in the variables Q and λ is

$$\begin{array}{ll} \text{maximize} & \lambda \\ \text{subject to} & (5.28), \qquad 5.35), \qquad \lambda I \leq Q \end{array}$$

• NORM-BOUND LDIS: For NLDIs, the EVP in the variables Q, μ and λ is

$$\begin{array}{ll} \text{maximize} & \lambda \\ \text{subject to} & (5.30), \quad (5.35), \quad \lambda I \le Q \end{array}$$

Remark: These results can be used as follows. The polytope $\mathcal{P}$ represents the allowable (or safe) operating region for the system. The ellipsoids found above can be interpreted as regions of safe initial conditions, i.e., initial conditions for which we can guarantee that the state always remains in the safe operating region.

5.2.3 Bound on return time

The *return time* of a stable LDI for the polytope $\mathcal{P}$ is defined as the smallest T such that if $x(0) \in \mathcal{P}$, then $x(t) \in \mathcal{P}$ for all $t \ge T$. Upper bounds on the return time can be found by solving EVPs.

• LTI SYSTEMS: Let us consider a positive decay rate α. If $Q > 0$ satisfies

$$QA^T + AQ + 2\alpha Q \le 0,$$

then $\mathcal{E}$ is an invariant ellipsoid, and moreover $x(0) \in \mathcal{E}$ implies that $x(t) \in e^{-\alpha t}\mathcal{E}$ for all $t \ge 0$. Therefore if T is such that

$$e^{-\alpha T}\mathcal{E} \subseteq \mathcal{P} \subseteq \mathcal{E},$$

we can conclude that if $x(0) \in \mathcal{P}$, then $x(t) \in \mathcal{P}$ for all $t \ge T$, so that T is an upper bound on the return time. If we use both representations of the polytope,

$$\mathcal{P} = \left\{ x \in \mathbf{R}^n \;\middle|\; a_k^T x \le 1, \quad k = 1, \ldots, q \right\} = \mathbf{Co}\{v_1, \ldots, v_p\},$$

the constraint $e^{-\alpha T}\mathcal{E} \subseteq \mathcal{P}$ is equivalent to the LMI in Q and γ

$$a_i^T Q a_i \le \gamma, \quad k = 1, \ldots, q, \tag{5.36}$$

where $\gamma = e^{2\alpha T}$, and the problem of finding the smallest such γ and therefore the smallest T (for a fixed α) is an EVP in the variables γ and Q:

$$\begin{array}{ll} \text{minimize} & \gamma \\ \text{subject to} & (5.19), \quad (5.31), \quad (5.36) \end{array}$$

• POLYTOPIC LDIS: For PLDIs, the smallest bound on the return time provable via invariant ellipsoids, with a given decay rate α, is obtained by solving the EVP in the variables Q and γ

$$\begin{array}{ll} \text{minimize} & \gamma \\ \text{subject to} & (5.21), \quad (5.31), \quad (5.36) \end{array}$$

• NORM-BOUND LDIS: For NLDIs, the smallest bound on the return time provable via invariant ellipsoids, with a given decay rate α, is obtained by solving the EVP in the variables Q, γ and μ

$$\begin{array}{ll} \text{minimize} & \gamma \\ \text{subject to} & (5.23), \quad (5.31), \quad (5.36) \end{array}$$

Notes and References

Quadratic stability for NLDIs

Lur'e and Postnikov [LP44] gave one of the earliest stability analyses for NLDIs: They considered the stability of the system

$$\dot{x} = Ax + b_p p, \qquad q = c_q x, \qquad p = \phi(q,t)q, \tag{5.37}$$

where p and q are scalar. This problem came to be popularly known as the problem of "absolute stability in automatic control". In the original setting of Lur'e and Postnikov, ϕ was assumed to be a time-invariant nonlinearity. Subsequently, various additional assumptions were made about ϕ, generating different special cases and solutions. (The family of systems of the form (5.37), where $\phi(q,t)$ can be any sector $[-1,1]$ nonlinearity, is an NLDI.) Pyatnitskii, in [PYA68], points out that by 1968, over 200 papers had been written about the system (5.37).

Among these, the ones most relevant to this book are undoubtedly from Yakubovich. As far as we know, Yakubovich was the first to make systematic use of LMIs along with the $\mathcal{S}$-procedure to prove stability of nonlinear control systems (see references [YAK62, YAK64, YAK67, YAK66, YAK77, YAK82]). The main idea of Yakubovich was to express the resulting LMIs as frequency-domain criteria for the transfer function $G(s) = c_q(sI - A)^{-1}b_p$. (Yakubovich calls this method the "method of matrix inequalities".) These criteria are most useful when dealing with experimental data arising from frequency response measurements; they are described in detail in the book by Narendra and Taylor [NT73]. See also [BAR70B]. Popov [POP73] and Willems [WIL71B, WIL74A] outline the relationship between the problem of absolute stability of automatic control and quadratic optimal control.

The case when p and q are vector-valued signals has been considered much more recently; see for instance, [ZK88, ZK87, KR88, BH89]. In [KPZ90, PD90], the authors remark that quadratic stability of NLDIs is equivalent to the $\mathbf{H}_\infty$ condition (5.13). However, as stated in [PD90], there is no fundamental difference between these and the older results of Yakubovich and Popov; the newer results can be regarded as extensions of the works of Yakubovich and Popov to feedback synthesis and to the case of "structured perturbations" [RCDP93, PZP+92].

While all the results presented in this chapter are based on Lyapunov stability techniques, they have very strong connections with the input-output approach to study the stability of nonlinear, uncertain systems. This approach was sparked by Popov [POP62], who used it to study the Lur'e system. Other major contributors include Zames [ZAM66A, ZAM66B, ZF67], Sandberg [SAN64, SAN65A, SAN65B] and Brockett [BRO65, BRO66, BW65]. See also the papers and the book by Willems [WIL69A, WIL69B, WIL71A]. The advantage of the input-output approach is that the systems are not required to have finite-dimensional state [DES65, DV75, LOG90]; the corresponding stability criteria are usually most easily expressed in the frequency domain, and result in *infinite-dimensional* convex optimization problems. Modern approaches from this viewpoint include Doyle's μ-analysis [DOY82] and Safonov's K_m-analysis [SAF82, SD83, SD84], implemented in the robustness analysis software packages μ-tools and K_m-tools [BDG+91, CS92A], which approximately solve these infinite-dimensional problems. See also [AS79, SS81].

Quadratic stability for PLDIs

This topic is much more recent than NLDIs, the likely reason being that the LMI expressing quadratic stability of a general PLDI (which is just a number of simultaneous Lyapunov inequalities (5.8)) cannot be converted to a Riccati inequality or a frequency-domain criterion. One of the first occurrences of PLDIs is in [HB76], where Horisberger and Belanger note that the problem of quadratic stability for a PLDI is convex (the authors write down the corresponding LMIs). For other results and computational procedures for PLDIs see [PS82, PS86, BY89, KP87A, KAM83, EZKN90, KB93, AGG94]; the article [KP87A], where the authors develop a subgradient method for proving quadratic stability of a PLDI, which is further refined in [MEI91]; the paper [PS86], where the discrete-time counterpart of this problem is considered. See also the articles by Gu et al. [GZL90, GL93B, GU94], Garofalo et al. [GCG93], and the survey article by Barmish et al. [BK93].

Necessary and sufficient conditions for LDI stability

Much attention has also focused on necessary and sufficient conditions for stability of LDIs (as opposed to quadratic stability); see for example [PR91B] or [MP86]. Earlier references on this topic include [PYA70B] which connects the problem of stability of an LDI to an optimal control problem, and [PYA70A, BT79, BT80], where the discrete-time counterpart of this problem is considered.

Vector Lyapunov functions

The technique of vector Lyapunov functions also yields search problems that can be expressed as LMIPs. One example is given in [CFA90]. Here we already have quadratic Lyapunov functions for a number of subsystems; the problem is to find an appropriate positive linear combination that proves, e.g., stability of an interconnected system.

Finally, let us point out that quadratic Lyapunov functions have been used to determine estimates of regions of stability for general nonlinear systems. See for example [GHA94, BD85].

Stable LDIs that are not quadratically stable

An LTI system is stable if and only if it is quadratically stable; this is just Lyapunov's stability theorem for linear systems (see e.g., [LYA47, P277]). It is possible, however, for an LDI to be stable without being quadratically stable. Here is a PLDI example:

$$\dot{x} = A(t)x, \qquad A(t) \in \mathbf{Co}\{A_1, A_2\},$$

$$A_1 = \begin{bmatrix} -100 & 0 \\ 0 & -1 \end{bmatrix}, \qquad A_2 = \begin{bmatrix} 8 & -9 \\ 120 & -18 \end{bmatrix}.$$

From (2.8), this PLDI is not quadratically stable if there exist $Q_0 \geq 0$, $Q_1 \geq 0$ and $Q_2 \geq 0$, not all zero, such that

$$Q_0 = A_1 Q_1 + Q_1 A_1^T + A_2 Q_2 + Q_2 A_2^T.$$

It can be verified that the matrices

$$Q_0 = \begin{bmatrix} 5.2 & 2 \\ 2 & 24 \end{bmatrix}, \ Q_1 = \begin{bmatrix} 0.1 & 3 \\ 3 & 90 \end{bmatrix}, \ Q_2 = \begin{bmatrix} 2.7 & 1 \\ 1 & 1 \end{bmatrix}$$

satisfy these duality conditions.

However, the piecewise quadratic Lyapunov function

$$V(x) = \max\left\{x^T P_1 x, x^T P_2 x\right\},$$
$$P_1 = \begin{bmatrix} 14 & -1 \\ -1 & 1 \end{bmatrix}, \ P_2 = \begin{bmatrix} 0 & 0 \\ 0 & 1 \end{bmatrix}, \qquad (5.38)$$

proves that the PLDI is stable. To show this, we use the $\mathcal{S}$-procedure. A necessary and sufficient condition for the Lyapunov function V defined in (5.38) to prove the stability of the PLDI is the existence of four nonnegative numbers λ_1, λ_2, λ_3, λ_4 such that

$$\begin{aligned}
&A_1^T P_1 + P_1 A_1 - \lambda_1 (P_2 - P_1) < 0, \\
&A_2^T P_1 + P_1 A_2 - \lambda_2 (P_2 - P_1) < 0, \\
&A_1^T P_2 + P_2 A_1 + \lambda_3 (P_2 - P_1) < 0, \;\; A_2^T P_2 + P_2 A_2 + \lambda_4 (P_2 - P_1) < 0.
\end{aligned} \tag{5.39}$$

It can be verified that $\lambda_1 = 50$, $\lambda_2 = 0$, $\lambda_3 = 1$, $\lambda_4 = 100$ are such numbers.

For other examples of stable LDIs that are not quadratically stable, see Brockett [BRO65, BRO77, BRO70] and Pyatnitskii [PYA71]. In [BRO77], Brockett uses the $\mathcal{S}$-procedure to prove that a certain piecewise quadratic form is a Lyapunov function and finds LMIs similar to (5.39); see also [VID78].

Finally, an LDI is stable if and only if there is a convex Lyapunov function that proves it. One such Lyapunov function is

$$V(\xi) = \sup \left\{ \|x(t)\| \mid x \text{ satisfies } (5.1),\; x(0) = \xi,\; t \geq 0 \right\}.$$

See Brayton and Tong [BT79, BT80], and also [ZUB55, MV63, DK71, VV85, PR91B, PR91A, RAP90, RAP93, MP89, MOL87]. In general, computing such Lyapunov functions is computationally intensive, if not intractable.

Nonlinear systems and fading memory

Consider the nonlinear system

$$\dot{x} = f(x, w, t), \tag{5.40}$$

with

$$\frac{\partial f}{\partial x} \in \Omega \text{ for all } x,\; w,\; t$$

where Ω is convex. Suppose the LDI $\dot{x} \in \Omega x$ is stable, i.e., all trajectories converge to zero as $t \to \infty$. This implies the system has *fading memory*, that is, for fixed input w, the difference between any two trajectories of (5.40) converges to zero. In other words, the system "forgets" its initial condition. (See [BC85].)

Fix an input w and let x and $\tilde{x}$ denote any two solutions of (5.40). Using the mean-value theorem from §4.3.1, we have

$$\frac{d}{dt}(x - \tilde{x}) = f(x, w, t) - f(\tilde{x}, w, t) = A(t)(x - \tilde{x})$$

for some $A(t) \in \Omega$. Since the LDI is stable, $x(t) - \tilde{x}(t)$ converges to zero as $t \to \infty$.

Relation between $\mathcal{S}$-procedure for DNLDIs and scaling

We discuss the interpretation of the diagonal matrix Λ in (5.15) as a *scaling matrix*.

Every trajectory of the DNLDI (5.3) is also a trajectory (and vice versa) of the DNLDI

$$\begin{aligned}
\dot{x} &= Ax + B_p T^{-1} p, \\
q &= T C_q x + T D_{qp} T^{-1} p, \\
p_i &= \delta_i(t) q_i, \qquad |\delta_i(t)| \leq 1, \qquad i = 1, \ldots, n_q
\end{aligned} \tag{5.41}$$

for any diagonal nonsingular T. Therefore, if DNLDI (5.41) is stable for some diagonal nonsingular T, then so is DNLDI (5.3). T is referred to as a scaling matrix.

Treating DNLDI (5.41) as an NLDI, and applying the condition (5.13) for quadratic stability, we require, for some diagonal nonsingular T,

$$P > 0, \qquad \begin{bmatrix} A^TP + PA + C_q^TT^2C_q & PB_pT^{-1} + C_q^TT^2D_{qp}T^{-1} \\ (PB_pT^{-1} + C_q^TT^2D_{qp}T^{-1})^T & -(I - T^{-1}D_{qp}^TT^2D_{qp}T^{-1}) \end{bmatrix} < 0,$$

An obvious congruence, followed by the substitution $T^2 = \Lambda$, yields the LMI condition in $P > 0$ and diagonal $\Lambda > 0$:

$$\begin{bmatrix} A^TP + PA + C_q^T\Lambda C_q & PB_p + C_q\Lambda D_{qp} \\ B_P^TP + D_{qp}^T\Lambda C_q & D_{qp}^T\Lambda D_q - \Lambda \end{bmatrix} < 0.$$

This is precisely LMI (5.15), the condition for stability of the DNLDI (5.3) obtained using the $\mathcal{S}$-procedure. This relation between scaling and the diagonal matrices arising from the $\mathcal{S}$-procedure is described in Boyd and Yang [BY89].

Pyatnitskii and Skorodinskii [PS82, PS83], and Kamenetskii [KAM83] reduce the problem of numerical search for appropriate scalings and the associated Lyapunov functions to a convex optimization problem. Saeki and Araki [SA88] also conclude convexity of the scaling problem and use the properties of M-matrices to obtain a solution.

Representing a DNLDI as a PLDI: Implications for quadratic stability

We saw in Chapter 4 that a DNLDI can be represented as a PLDI, where the number L of vertices can be as large as 2^{n_q}. From the results of §5.1, checking quadratic stability for this PLDI requires the solution of L simultaneous Lyapunov inequalities in $P > 0$ (LMI (5.8)). If the system is represented as a DNLDI, a sufficient condition for quadratic stability is given by the LMI (5.15), in the variables $P > 0$ and $\Lambda > 0$.

Comparing the two conditions for quadratic stability, we observe that while the quadratic stability condition (5.8) for LDI (5.3), obtained by representing it as a PLDI, is both necessary and sufficient (i.e., not conservative), the size of the LMI grows *exponentially* with n_q, the size of Δ. On the other hand, the size of the LMI (5.15) grows *polynomially* with n_q, but this LMI yields only a sufficient condition for quadratic stability of LDI (5.3). Discussion of this issue can be found in Kamenetskii [KAM83] and Pyatnitskii and Skorodinskii [PS82].

Chapter 6

Analysis of LDIs: Input/Output Properties

6.1 Input-to-State Properties

We first consider input-to-state properties of the LDI

$$\dot{x} = A(t)x + B_w(t)w, \qquad [A(t) \quad B_w(t)] \in \Omega, \tag{6.1}$$

where w is an exogenous input signal.

- LTI SYSTEMS: The system has the form $\dot{x} = Ax + B_w w$.
- POLYTOPIC LDIS: Here the description is

$$\dot{x} = A(t)x + B_w(t)w, \qquad [A(t) \quad B_w(t)] \in \mathbf{Co}\left\{[A_1 \quad B_{w,1}], \ldots, [A_L \quad B_{w,L}]\right\}.$$

- NORM-BOUND LDIS: NLDIs have the form

$$\dot{x} = Ax + B_p p + B_w w, \qquad q = C_q x, \qquad p = \Delta(t) q, \qquad \|\Delta(t)\| \le 1,$$

or equivalently,

$$\dot{x} = Ax + B_p p + B_w w, \qquad q = C_q x, \qquad p^T p \le q^T q.$$

- DIAGONAL NORM-BOUND LDIS: DNLDIs have the form

$$\begin{aligned} &\dot{x} = Ax + B_p p + B_w w, \qquad q = C_q x, \\ &p_i = \delta_i(t) q_i, \qquad |\delta_i(t)| \le 1, \quad i = 1, \ldots, n_q, \end{aligned}$$

or equivalently,

$$\dot{x} = Ax + B_p p + B_w w, \qquad q = C_q x, \qquad |p_i| \le |q_i|, \quad i = 1, \ldots, n_q.$$

6.1.1 Reachable sets with unit-energy inputs

Let $\mathcal{R}_{\rm ue}$ denote the set of reachable states with unit-energy inputs for the LDI (6.1), i.e.,

$$\mathcal{R}_{\rm ue} \stackrel{\Delta}{=} \left\{ x(T) \;\middle|\; \begin{array}{ll} x,\, w \text{ satisfy (6.1)}, & x(0) = 0 \\ \displaystyle\int_0^T w^T w \, dt \le 1, & T \ge 0 \end{array} \right\}.$$

We will bound $\mathcal{R}_{\rm ue}$ by ellipsoids of the form

$$\mathcal{E} = \left\{\xi \mid \xi^T P \xi \le 1\right\}, \tag{6.2}$$

where $P > 0$.

Suppose that the function $V(\xi) = \xi^T P \xi$, with $P > 0$, satisfies

$$dV(x)/dt \le w^T w \quad \text{for all } x,\ w \text{ satisfying (6.1).} \tag{6.3}$$

Integrating both sides from 0 to T, we get

$$V(x(T)) - V(x(0)) \le \int_0^T w^T w\, dt.$$

Noting that $V(x(0)) = 0$ since $x(0) = 0$, we get

$$V(x(T)) \le \int_0^T w^T w\, dt \le 1,$$

for every $T \ge 0$, and every input w such that $\int_0^T w^T w\, dt \le 1$. In other words, the ellipsoid $\mathcal{E}$ contains the reachable set $\mathcal{R}_{\rm ue}$.

- LTI SYSTEMS: Condition (6.3) is equivalent to the LMI in P

$$P > 0, \qquad \begin{bmatrix} A^T P + PA & PB_w \\ B_w^T P & -I \end{bmatrix} \le 0, \tag{6.4}$$

which can also be expressed as an LMI in $Q = P^{-1}$

$$Q > 0, \qquad AQ + QA^T + B_w B_w^T \le 0. \tag{6.5}$$

Remark: For a controllable LTI system, the reachable set is the ellipsoid $\{\xi \mid \xi^T W_{\rm c}^{-1} \xi \le 1\}$, where $W_{\rm c}$ is the *controllability Gramian*, defined by

$$W_{\rm c} \triangleq \int_0^\infty e^{At} B_w B_w^T e^{A^T t}\, dt.$$

Since $W_{\rm c}$ satisfies the Lyapunov equation

$$AW_{\rm c} + W_{\rm c} A^T + B_w B_w^T = 0, \tag{6.6}$$

we see that $Q = W_{\rm c}$ satisfies (6.5). Thus the ellipsoidal bound is sharp for LTI systems.

- POLYTOPIC LDIS: For PLDIs, condition (6.3) holds if and only if

$$P > 0 \text{ and } \begin{bmatrix} A(t)^T P + PA(t) & PB_w(t) \\ B_w(t)^T P & -I \end{bmatrix} \le 0 \text{ for all } t \ge 0. \tag{6.7}$$

Inequality (6.7) holds if and only if

$$P > 0, \qquad \begin{bmatrix} A_i^T P + PA_i & PB_{w,i} \\ B_{w,i}^T P & -I \end{bmatrix} \le 0, \quad i = 1, \ldots, L. \tag{6.8}$$

Thus the ellipsoid $\mathcal{E}$ contains $\mathcal{R}_{\rm ue}$ if the LMI (6.8) holds.

Alternatively, we can write (6.8) as

$$Q > 0, \qquad QA_i^T + A_iQ + B_{w,i}B_{w,i}^T \le 0, \quad i = 1, \ldots, L, \tag{6.9}$$

where $Q = P^{-1}$. Note that this condition implies that $\mathcal{E}$ is invariant.

- NORM-BOUND LDIS: Condition (6.3) holds if and only if $P > 0$ and

$$\xi^T(A^TP + PA)\xi + 2\xi^TP(B_p\pi + B_w\omega) - \omega^T\omega \le 0$$

hold for every ω and for every ξ and π satisfying

$$\pi^T\pi - \xi^TC_q^TC_q\xi \le 0.$$

Using the $\mathcal{S}$-procedure, this is equivalent to the existence of P and λ satisfying

$$P > 0, \qquad \lambda \ge 0, \qquad \begin{bmatrix} A^TP + PA + \lambda C_q^TC_q & PB_p & PB_w \\ B_p^TP & -\lambda I & 0 \\ B_w^TP & 0 & -I \end{bmatrix} \le 0. \tag{6.10}$$

Equivalently, defining $Q = P^{-1}$, condition (6.3) holds if and only if there exist Q and μ satisfying

$$Q > 0, \qquad \mu \ge 0, \qquad \begin{bmatrix} AQ + QA^T + B_wB_w^T + \mu B_pB_p^T & QC_q^T \\ C_qQ & -\mu I \end{bmatrix} \le 0. \tag{6.11}$$

- DIAGONAL NORM-BOUND LDIS: Condition (6.3) holds if and only if $P > 0$ and for all ξ and π that satisfy

$$\pi_i^2 \le (C_{q,i}\xi)^T(C_{q,i}\xi), \qquad i = 1, \ldots, n_q,$$

we have for all ω,

$$\xi^T(A^TP + PA)\xi + 2\xi^TP(B_p\pi + B_w\omega) - \omega^T\omega \le 0.$$

It follows from the $\mathcal{S}$-procedure that this condition holds if there exist P and $\Lambda = \mathbf{diag}(\lambda, \ldots, \lambda_{n_q})$ satisfying

$$P > 0, \qquad \Lambda \ge 0, \qquad \begin{bmatrix} A^TP + PA + C_q^T\Lambda C_q & PB_p & PB_w \\ B_p^TP & -\Lambda & 0 \\ B_w^TP & 0 & -I \end{bmatrix} \le 0.$$

Equivalently, defining $Q = P^{-1}$, condition (6.3) holds if there exist Q and a diagonal matrix M such that

$$Q > 0, \qquad M \ge 0, \qquad \begin{bmatrix} AQ + QA^T + B_wB_w^T + B_pMB_p^T & QC_q^T \\ C_qQ & -M \end{bmatrix} \le 0. \tag{6.12}$$

For the remainder of this section we leave the extension to DNLDIs to the reader.

The results described above give us a set of ellipsoidal outer approximations of $\mathcal{R}_{\mathrm{ue}}$. We can optimize over this set in several ways.

Smallest outer approximations

• LTI SYSTEMS: For LTI systems, we can minimize the volume over ellipsoids $\mathcal{E}$ with P satisfying (6.3) by minimizing $\log\det P^{-1}$ over the variable P subject to (6.4). This is a CP.

If (A, B_w) is uncontrollable, the volume of the reachable set is zero, and this CP will be unbounded below. If (A, B_w) is controllable, then the CP finds the exact reachable set, which corresponds to $P = W_c^{-1}$.

• POLYTOPIC LDIS: For PLDIs, the minimum volume ellipsoid of the form (6.2) with P satisfying (6.3), is found by minimizing $\log\det P^{-1}$ subject to (6.8). This is a CP.

• NORM-BOUND LDIS: For NLDIs, the minimum volume ellipsoid of the form (6.2), satisfying (6.3), is obtained by minimizing $\log\det P^{-1}$ over the variables P and λ subject to (6.10). This is again a CP.

We can also minimize the diameter of ellipsoids of the form (6.2), satisfying (6.3), by replacing the objective function $\log\det P^{-1}$ in the CPs by the objective function $\lambda_{\max}(P^{-1})$. This yields EVPs.

Testing if a point is outside the reachable set

The point x_0 lies outside the reachable set if there exists an ellipsoidal outer approximation of $\mathcal{R}_{\rm ue}$ that does not contain it. For LTI systems, x_0 does not belong to the reachable set if there exists P satisfying $x_0^T P x_0 > 1$ and (6.4). This is an LMIP. Of course, if (A, B_w) is controllable, then x_0 does not belong to the reachable set if and only if $x_0^T W_c^{-1} x_0 > 1$, where W_c is defined by the equation (6.6).

For PLDIs, a sufficient condition for a point x_0 to lie outside the reachable set $\mathcal{R}_{\rm ue}$ can be checked via an LMIP in the variable P: $x_0^T P x_0 > 1$ and (6.8). For NLDIs, x_0 lies outside the reachable set $\mathcal{R}_{\rm ue}$ if there exist P and λ satisfying $x_0^T P x_0 > 1$ and (6.10), another LMIP.

Testing if the reachable set lies outside a half-space

The reachable set $\mathcal{R}_{\rm ue}$ lies outside the half-space $\mathcal{H} = \left\{\xi \in \mathbf{R}^n \;\middle|\; a^T\xi > 1\right\}$ if and only if an ellipsoid $\mathcal{E}$ containing $\mathcal{R}_{\rm ue}$ lies outside $\mathcal{H}$, that is, the minimum value of $\xi^T Q^{-1}\xi$ over ξ satisfying $a^T\xi > 1$ exceeds one. We easily see that

$$\min\left\{\xi^T Q^{-1}\xi \;\middle|\; a^T\xi > 1\right\} = 1/(a^T Q a).$$

Therefore, for LTI systems, $\mathcal{H}$ does not intersect the reachable set if there exists Q satisfying $a^T Q a < 1$ and (6.5). This is an LMIP. If (A, B_w) is controllable, then $\mathcal{H}$ does not intersect the reachable set if and only if $a^T W_c a < 1$, where W_c is defined by the equation (6.6).

For PLDIs, a sufficient condition for $\mathcal{H}$ not to intersect the reachable set $\mathcal{R}_{\rm ue}$ can be checked via an LMIP in the variable Q: $a^T Q a < 1$ and (6.9). For NLDIs, $\mathcal{H}$ does not intersect the reachable set $\mathcal{R}_{\rm ue}$ if there exists Q and μ satisfying $a^T Q a < 1$ and (6.11). This is an LMIP.

Testing if the reachable set is contained in a polytope

Since a polytope can be expressed as an intersection of half-spaces (these determine its "faces"), we can immediately use the results of the previous subsection in order to check if the reachable set is contained in a polytope. (Note we only obtain sufficient conditions.) We remark that we can use different ellipsoidal outer approximations to check different faces.

6.1.2 Reachable sets with componentwise unit-energy inputs

We now turn to a variation of the previous problem; we consider the set of reachable states with inputs whose components have unit-energy, that is, we consider the set

$$\mathcal{R}_{\mathrm{uce}} \triangleq \left\{ x(T) \;\middle|\; \begin{array}{lll} x,\, w \text{ satisfy (6.1)}, & x(0)=0, & \\ \int_0^T w_i^2\, dt \le 1, & i=1,\dots,n_w, & T \ge 0 \end{array} \right\}.$$

Suppose there is a positive-definite quadratic function $V(\xi) = \xi^T P \xi$, and $R = \mathbf{diag}(r_1, \dots, r_{n_w})$ such that

$$\begin{array}{l} P > 0, \qquad R \ge 0, \qquad \mathbf{Tr}\, R = 1, \\ \dfrac{d}{dt} V(x) \le w^T R w \text{ for all } x \text{ and } w \text{ satisfying (6.1).} \end{array} \tag{6.13}$$

Then the ellipsoid $\mathcal{E}$ given by (6.2) contains the reachable set. To prove this, we integrate both sides of the last inequality from 0 to T, to get

$$V(x(T)) - V(x(0)) \le \int_0^T w^T R w \, dt.$$

Noting that $V(x(0)) = 0$ since $x(0) = 0$, we get

$$V(x(T)) \le \int_0^T w^T R w \, dt \le 1.$$

Let us now consider condition (6.13) for the various LDIs:

• LTI SYSTEMS: For LTI systems, condition (6.13) is equivalent to the LMI in P and R:

$$\begin{array}{l} P > 0,\ R \ge 0 \text{ and diagonal},\ \mathbf{Tr}\, R = 1, \\ \left[\begin{array}{cc} A^T P + PA & PB_w \\ B_w^T P & -R \end{array} \right] \le 0. \end{array} \tag{6.14}$$

Therefore, if there exist P and R such that (6.14) is satisfied, then the ellipsoid $\mathcal{E}$ contains the reachable set $\mathcal{R}_{\mathrm{uce}}$.

Alternatively, using the variable $Q = P^{-1}$, we can write an equivalent LMI in Q and R:

$$\begin{array}{l} Q > 0,\ R \ge 0 \text{ and diagonal},\ \mathbf{Tr}\, R = 1, \\ \left[\begin{array}{cc} QA^T + AQ & B_w \\ B_w^T & -R \end{array} \right] \le 0. \end{array} \tag{6.15}$$

Remark: It turns out that with LTI systems, the reachable set is equal to the intersection of all such ellipsoids. (For more details on this, see the Notes and References.)

• POLYTOPIC LDIs: For PLDIs, condition (6.13) holds if and only if

$$P > 0,\ R \geq 0 \text{ and diagonal, } \mathbf{Tr}\, R = 1,$$
$$\begin{bmatrix} A_i^T P + PA_i & PB_{w,i} \\ B_{w,i}^T P & -R \end{bmatrix} \leq 0, \quad i = 1, \ldots, L. \tag{6.16}$$

Therefore, if there exist P and R such that (6.16) holds, then the ellipsoid $\mathcal{E} = \{\xi \mid \xi^T P \xi \leq 1\}$ contains $\mathcal{R}_{\rm uce}$.

With $Q = P^{-1}$, condition (6.13) is equivalent to

$$Q > 0,\ R \geq 0 \text{ and diagonal, } \mathbf{Tr}\, R = 1,$$
$$\begin{bmatrix} QA_i^T + A_i Q & B_{w,i} \\ B_{w,i}^T & -R \end{bmatrix} \leq 0, \quad i = 1, \ldots, L. \tag{6.17}$$

• NORM-BOUND LDIs: For NLDIs, applying the $\mathcal{S}$-procedure, condition (6.13) holds if and only if there exist P, λ and R such that

$$P > 0,\ R \geq 0 \text{ and diagonal, } \mathbf{Tr}\, R = 1,\ \lambda \geq 0,$$
$$\begin{bmatrix} A^T P + PA + \lambda C_q^T C_q & PB_p & PB_w \\ B_p^T P & -\lambda I & 0 \\ B_w^T P & 0 & -R \end{bmatrix} \leq 0. \tag{6.18}$$

Therefore, if there exists P, λ and R such that (6.18) is satisfied, then the ellipsoid $\mathcal{E}$ contains the reachable set $\mathcal{R}_{\rm uce}$.

With $Q = P^{-1}$, condition (6.13) is also equivalent to the existence of Q, R and μ such that

$$Q > 0,\ R \geq 0 \text{ and diagonal, } \mathbf{Tr}\, R = 1,\ \mu \geq 0,$$
$$\begin{bmatrix} AQ + QA^T + \mu B_p B_p^T & QC_q^T & B_w \\ C_q Q & -\mu I & 0 \\ B_w^T & 0 & -R \end{bmatrix} \leq 0. \tag{6.19}$$

As in the previous section, we can find the minimum volume and minimum diameter ellipsoids among all ellipsoids of the form (6.2), satisfying condition (6.13), by solving CPs and EVPs, respectively. We can also check that a point does not belong to the reachable set or that a half-space does not intersect the reachable set, by solving appropriate LMIPs.

6.1.3 Reachable sets with unit-peak inputs

We consider reachable sets with inputs w that satisfy $w^T w \leq 1$. Thus, we are interested in the set

$$\mathcal{R}_{\rm up} \stackrel{\Delta}{=} \left\{ x(T) \;\middle|\; \begin{array}{ll} x,\, w \text{ satisfy (6.1)}, & x(0) = 0, \\ w^T w \leq 1, & T \geq 0 \end{array} \right\}.$$

Suppose that there exists a quadratic function $V(\xi) = \xi^T P \xi$ with

$$\begin{array}{l} P > 0, \text{ and } dV(x)/dt \leq 0 \text{ for all } x,\, w \\ \text{satisfying (6.1), } w^T w \leq 1 \text{ and } V(x) \geq 1. \end{array} \tag{6.20}$$

Then, the ellipsoid $\mathcal{E}$ given by (6.2) contains the reachable set $\mathcal{R}_{\rm up}$. Let us consider now condition (6.20) for the various LDIs:

• LTI SYSTEMS: For LTI systems, condition (6.20) is equivalent to $P > 0$ and

$$\xi^T(A^TP + PA)\xi + \xi^T PB_w w + w^T B_w^T P\xi \leq 0$$

for any ξ and w satisfying

$$w^Tw \leq 1 \qquad \text{and} \qquad \xi^T P\xi \geq 1.$$

Using the $\mathcal{S}$-procedure, we conclude that condition (6.20) holds if there exist $\alpha \geq 0$ and $\beta \geq 0$ such that for all x and w,

$$\begin{bmatrix} x \\ w \end{bmatrix}^T \begin{bmatrix} A^TP + PA + \alpha P & PB_w \\ B_w^T P & -\beta I \end{bmatrix} \begin{bmatrix} x \\ w \end{bmatrix} + \beta - \alpha \leq 0, \tag{6.21}$$

or equivalently

$$\begin{bmatrix} A^TP + PA + \alpha P & PB_w & 0 \\ B_w^T P & -\beta I & 0 \\ 0 & 0 & \beta - \alpha \end{bmatrix} \leq 0. \tag{6.22}$$

Clearly we must have $\alpha \geq \beta$. Next, if (6.22) holds for some (α_0, β_0), then it holds for all $\alpha_0 \geq \beta \geq \beta_0$. Therefore, we can assume without loss of generality that $\beta = \alpha$, and rewrite (6.21) as

$$\begin{bmatrix} A^TP + PA + \alpha P & PB_w \\ B_w^T P & -\alpha I \end{bmatrix} \leq 0. \tag{6.23}$$

Therefore, if there exists P and α satisfying

$$P > 0, \qquad \alpha \geq 0, \qquad \text{and (6.23)}, \tag{6.24}$$

then the ellipsoid $\mathcal{E}$ contains the reachable set $\mathcal{R}_{\rm up}$. Note that inequality (6.23) is not an LMI in α and P; however, for fixed α it is an LMI in P.

Alternatively, condition (6.23) is equivalent to the following inequality in $Q = P^{-1}$ and α:

$$\begin{bmatrix} QA^T + AQ + \alpha Q & B_w \\ B_w^T & -\alpha I \end{bmatrix} \leq 0. \tag{6.25}$$

• POLYTOPIC LDIS: Condition (6.20) holds if $P > 0$ and there exists $\alpha \geq 0$ satisfying

$$\begin{bmatrix} A_i^TP + PA_i + \alpha P & PB_{w,i} \\ B_{w,i}^T P & -\alpha I \end{bmatrix} \leq 0, \quad i = 1, \ldots, L. \tag{6.26}$$

Therefore, if there exists P and α satisfying

$$P > 0, \qquad \alpha \geq 0, \qquad \text{and (6.26)}, \tag{6.27}$$

then the ellipsoid $\mathcal{E}$ contains the reachable set $\mathcal{R}_{\rm up}$. Once again, we note that condition (6.26) is not an LMI in α and P, but is an LMI in P for fixed α. We can also rewrite condition (6.26) in terms of $Q = P^{-1}$ as the equivalent inequality

$$\begin{bmatrix} QA_i^T + A_iQ + \alpha Q & B_{w,i} \\ B_{w,i}^T & -\alpha I \end{bmatrix} \leq 0, \quad i = 1, \ldots, L. \tag{6.28}$$

• NORM-BOUND LDIS: For NLDIs, condition (6.20) holds if and only if $P > 0$ and

$$\xi^T(A^TP + PA)\xi + 2\xi^TP(B_w\omega + B_p\pi) \leq 0$$

for every ω, ξ and π satisfying

$$\omega^T\omega \leq 1, \qquad \xi^TP\xi \geq 1, \qquad \pi^T\pi - \xi^TC_q^TC_q\xi \leq 0.$$

Therefore, using the $\mathcal{S}$-procedure, a sufficient condition for (6.20) to hold is that there exist $\alpha \geq 0$ and $\lambda \geq 0$ such that

$$\begin{bmatrix} A^TP + PA + \alpha P + \lambda C_q^TC_q & PB_p & PB_w \\ B_p^TP & -\lambda I & 0 \\ B_w^TP & 0 & -\alpha I \end{bmatrix} \leq 0. \tag{6.29}$$

If there exists P, α and λ satisfying

$$P > 0, \qquad \alpha \geq 0, \qquad \lambda \geq 0, \qquad \text{and (6.29)}, \tag{6.30}$$

the ellipsoid $\mathcal{E}_{P^{-1}}$ contains the reachable set $\mathcal{R}_{\rm up}$. We note again that inequality (6.29) is not an LMI in P and α, but is an LMI in P for fixed α.

Defining $Q = P^{-1}$, condition (6.20) is also implied by the existence of $\mu \geq 0$ and $\alpha \geq 0$ such that

$$\begin{bmatrix} QA^T + AQ + \alpha Q + \mu B_p^TB_p & QC_q^T & B_w \\ C_qQ & -\mu I & 0 \\ B_w^T & 0 & -\alpha I \end{bmatrix} \leq 0. \tag{6.31}$$

(In fact, the existence of $\lambda \geq 0$ and $\alpha \geq 0$ satisfying (6.29) is equivalent to the existence of $\mu \geq 0$ and $\alpha \geq 0$ satisfying (6.31).)

For fixed α, we can minimize the volume or diameter among all ellipsoids $\mathcal{E}$ satisfying condition (6.24) for LTI systems, condition (6.27) for PLDIs and condition (6.30) for NLDIs by forming the appropriate CPs, or EVPs, respectively.

> **Remark**: Inequality (6.23) can also be derived by combining the results of §6.1.1 with an *exponential time-weighting* of the input w. See the Notes and References for details.
> As in §6.1.2, it is possible to determine outer ellipsoidal approximations of the reachable set of LTI systems, PLDIs and NLDIs, subject to componentwise unit-peak inputs.

6.2 State-to-Output Properties

We now consider state-to-output properties for the LDI

$$\dot{x} = A(t)x, \qquad z = C_z(t)x \tag{6.32}$$

where

$$\begin{bmatrix} A(t) \\ C_z(t) \end{bmatrix} \in \Omega. \tag{6.33}$$

• LTI SYSTEMS: LTI systems have the form $\dot{x} = Ax$, $z = C_zx$.

• POLYTOPIC LDIS: PLDIs have the form $\dot{x} = A(t)x$, $z = C_z(t)x$, where

$$\begin{bmatrix} A(t) \\ C_z(t) \end{bmatrix} \in \mathbf{Co}\left\{ \begin{bmatrix} A_1 \\ C_1 \end{bmatrix}, \ldots, \begin{bmatrix} A_L \\ C_L \end{bmatrix} \right\}.$$

• NORM-BOUND LDIS: NLDIs have the form

$$\dot{x} = Ax + B_p p, \qquad q = C_q x, \qquad z = C_z x \qquad p = \Delta(t) q, \qquad \|\Delta(t)\| \le 1,$$

or equivalently

$$\dot{x} = Ax + B_p p, \qquad q = C_q x, \qquad z = C_z x, \qquad p^T p \le q^T q.$$

• DIAGONAL NORM-BOUND LDIS: DNLDIs have the form

$$\begin{array}{ll} \dot{x} = Ax + B_p p, & q = C_q x, \\ z = C_z x, & p_i = \delta_i(t) q_i, \qquad |\delta_i(t)| \le 1, \quad i = 1, \ldots, n_q, \end{array}$$

or equivalently

$$\begin{array}{ll} \dot{x} = Ax + B_p p, & q = C_q x, \\ z = C_z x, & |p_i| \le |q_i|, \qquad i = 1, \ldots, n_q. \end{array}$$

6.2.1 Bounds on output energy

We seek the maximum output energy given a certain initial state,

$$\max \left\{ \int_0^\infty z^T z \, dt \;\middle|\; \dot{x} = A(t)x, \quad z = C_z(t)x \right\}, \tag{6.34}$$

where $x(0)$ is given, and the maximum is taken over $A(t)$, $C_z(t)$ such that (6.33) holds.

Suppose there exists a quadratic function $V(\xi) = \xi^T P \xi$ such that

$$P > 0 \text{ and } \frac{d}{dt} V(x) \le -z^T z, \quad \text{for every } x \text{ and } z \text{ satisfying (6.32).} \tag{6.35}$$

Then, integrating both sides of the second inequality in (6.35) from 0 to T, we get

$$V(x(T)) - V(x(0)) \le -\int_0^T z^T z \, dt.$$

for every $T \ge 0$. Since $V(x(T)) \ge 0$, we conclude that $V(x(0)) = x(0)^T P x(0)$ is an upper bound on the maximum energy of the output z given the initial condition $x(0)$.

We now derive LMIs that provide upper bounds on the output energy for the various LDIs.

• LTI SYSTEMS: In the case of LTI systems, condition (6.35) is equivalent to

$$P > 0, \qquad A^T P + PA + C_z^T C_z \le 0. \tag{6.36}$$

Therefore, we obtain the best upper bound on the output energy provable via quadratic functions by solving the EVP

$$\begin{array}{ll} \text{minimize} & x(0)^T P x(0) \\ \text{subject to} & P > 0, \qquad (6.36) \end{array} \tag{6.37}$$

In this case the solution can be found analytically, and it is exactly equal to the output energy, which is $x(0)^T W_o x(0)$, where W_o is the *observability Gramian* of the system, defined by

$$W_o \triangleq \int_0^\infty e^{A^T t} C_z^T C_z e^{At}\, dt.$$

Since W_o satisfies the Lyapunov equation

$$A^T W_o + W_o A + C_z^T C_z = 0, \tag{6.38}$$

it satisfies (6.35).

With $Q = P^{-1}$, condition (6.35) is equivalent to

$$Q > 0, \qquad \begin{bmatrix} AQ + QA^T & QC_z^T \\ C_z Q & -I \end{bmatrix} \leq 0. \tag{6.39}$$

If Q satisfies (6.39), then an upper bound on the output energy (6.34) is $x(0)^T Q^{-1} x(0)$.

• POLYTOPIC LDIS: For PLDIs, condition (6.35) is equivalent to

$$P > 0, \qquad A(t)^T P + PA(t) + C_z(t)^T C_z(t) \leq 0 \qquad \text{for all } t \geq 0. \tag{6.40}$$

Inequalities (6.40) hold if and only if the following LMI in the variable P holds:

$$P > 0, \qquad A_i^T P + PA_i + C_{z,i}^T C_{z,i} \leq 0, \quad i = 1, \ldots, L. \tag{6.41}$$

Therefore the EVP corresponding to finding the best upper bound on the output energy provable via quadratic functions is

$$\begin{array}{ll} \text{minimize} & x(0)^T P x(0) \\ \text{subject to} & P > 0, \qquad (6.41) \end{array} \tag{6.42}$$

Defining $Q = P^{-1}$, condition (6.35) is also equivalent to the LMI in Q

$$Q > 0, \qquad \begin{bmatrix} A_i Q + QA_i^T & QC_{z,i}^T \\ C_{z,i} Q & -I \end{bmatrix} \leq 0.$$

• NORM-BOUND LDIS: For NLDIs, condition (6.35) is equivalent to $P > 0$ and

$$\xi^T (A^T P + PA + C_z^T C_z) \xi + 2\xi^T P B_p \pi \leq 0$$

for every ξ and π that satisfy $\pi^T \pi \leq \xi^T C_q^T C_q \xi$. Using the $\mathcal{S}$-procedure, an equivalent condition is the existence of P and λ such that

$$\begin{array}{l} P > 0, \qquad \lambda \geq 0, \\ \begin{bmatrix} A^T P + PA + C_z^T C_z + \lambda C_q^T C_q & PB_p \\ B_p^T P & -\lambda I \end{bmatrix} \leq 0. \end{array} \tag{6.43}$$

Therefore we obtain the smallest upper bound on the output energy (6.34) provable via quadratic functions satisfying condition (6.35) by solving the following EVP in the variables P and λ

$$\begin{array}{ll} \text{minimize} & x(0)^T P x(0) \\ \text{subject to} & P > 0, \qquad \lambda \geq 0, \qquad (6.43) \end{array} \tag{6.44}$$

With $Q = P^{-1}$, condition (6.35) is also equivalent to the existence of $\mu \geq 0$ satisfying

$$Q > 0, \qquad \begin{bmatrix} AQ + QA^T + \mu B_p B_p^T & QC_z^T & QC_q^T \\ C_z Q & -I & 0 \\ C_q Q & 0 & -\mu I \end{bmatrix} \leq 0.$$

- DIAGONAL NORM-BOUND LDIS: For DNLDIs, condition (6.35) is equivalent to

$$\xi^T (A^T P + PA + C_z^T C_z)\xi + 2\xi^T P B_p \pi \leq 0,$$

for every ξ and π that satisfy

$$\pi_i^2 \leq (C_{q,i}\xi)^T (C_{q,i}\xi), \quad i = 1, \ldots, n_q.$$

Using the $\mathcal{S}$-procedure, a sufficient condition is that there exists a diagonal $\Lambda \geq 0$ such that

$$\begin{bmatrix} A^T P + PA + C_z^T C_z + C_q^T \Lambda C_q & PB_p \\ B_p^T P & -\Lambda \end{bmatrix} \leq 0. \tag{6.45}$$

Therefore we obtain the smallest upper bound on the output energy (6.34) provable via quadratic functions satisfying condition (6.35) and the $\mathcal{S}$-procedure by solving the following EVP in the variables P and Λ:

$$\begin{array}{ll} \text{minimize} & x(0)^T P x(0) \\ \text{subject to} & P > 0, \quad \Lambda \geq 0 \text{ and diagonal}, \quad (6.45) \end{array}$$

With $Q = P^{-1}$, condition (6.35) is also equivalent to the existence of a diagonal matrix $M \geq 0$ satisfying

$$\begin{bmatrix} AQ + QA^T + B_p M B_p^T & QC_z^T & QC_q^T \\ C_z Q & -I & 0 \\ C_q Q & 0 & -M \end{bmatrix} \leq 0.$$

Once again, we leave it to the reader to extend the results in the remainder of the section to DNLDIs.

Maximum output energy extractable from a set

As an extension, we seek bounds on the maximum extractable energy from $x(0)$, which is known only to lie in a polytope $\mathcal{P} = \mathbf{Co}\{v_1, \ldots, v_p\}$.

We first note that since $P > 0$, $x(0)^T P x(0)$ takes its maximum value on one of the vertices $v_1, \ldots, v_p$. Therefore, the smallest upper bound on the extractable energy from $\mathcal{P}$ provable via quadratic Lyapunov functions for any of our LDIs is computed by solving (6.37), (6.42) or (6.44) p times, successively setting $x(0) = v_i$, $i = 1, \ldots, p$. The maximum of the p resulting optimal values is the smallest upper bound sought.

As a variation, suppose that $x(0)$ is only known to belong to the ellipsoid $\mathcal{E} = \{\xi \mid \xi^T X \xi \leq 1\}$. Then, the smallest upper bound on the maximum output energy extractable from $\mathcal{E}$ provable via quadratic functions satisfying (6.35) is obtained by replacing the objective function $x(0)^T P x(0)$ in the EVPs (6.37), (6.42), and (6.44) by the objective function $\lambda_{\max}(X^{-1/2} P X^{-1/2})$, and solving the corresponding EVPs. For LTI systems, the exact value of the maximum extractable output energy is obtained.

Finally, if x_0 is a random variable with $\mathbf{E}\, x_0 x_0^T = X_0$, then we obtain the smallest upper bound on the expected output energy provable via quadratic functions by replacing the objective function $x(0)^T P x(0)$ in the EVPs (6.37), (6.42), and (6.44) by the objective function $\mathbf{Tr}\, X_0 P$ and solving the corresponding EVPs.

6.2.2 Bounds on output peak

It is possible to derive bounds on $\|z(t)\|$ using invariant ellipsoids.

Assume first that the initial condition $x(0)$ is known. Suppose $\mathcal{E} = \{\xi \mid \xi^T P \xi \le 1\}$ is an invariant ellipsoid containing $x(0)$ for the LDI (6.32). Then

$$z(t)^T z(t) \le \max_{\xi \in \mathcal{E}} \xi^T C_z(t)^T C_z(t) \xi$$

for all $t \ge 0$. We can express $\max_{\xi \in \mathcal{E}} \xi^T C_z(t)^T C_z(t) \xi$ as the square root of the minimum of δ subject to

$$\begin{bmatrix} P & C_z^T \\ C_z & \delta I \end{bmatrix} \ge 0. \tag{6.46}$$

• LTI SYSTEMS: The smallest bound on the output peak that can be obtained via invariant ellipsoids is the square root of the optimal value of the EVP in the variables P and δ

$$\begin{array}{ll} \text{minimize} & \delta \\ \text{subject to} & P > 0, \quad x(0)^T P x(0) \le 1, \\ & (6.46), \quad A^T P + PA \le 0 \end{array} \tag{6.47}$$

• POLYTOPIC LDIs: For PLDIs, we obtain the smallest upper bound on $\|z\|$ provable via invariant ellipsoids by taking the square root of the optimal value of the following EVP in the variables P and δ

$$\begin{array}{ll} \text{minimize} & \delta \\ \text{subject to} & P > 0, \quad x(0)^T P x(0) \le 1, \quad (6.46), \\ & A_i^T P + PA_i \le 0, \quad i = 1, \ldots, L \end{array}$$

• NORM-BOUND LDIs: For NLDIs, we obtain the smallest upper bound on $\|z\|$ provable via invariant ellipsoids by taking the square root of the optimal value of the following EVP in the variables P, λ and δ

$$\begin{array}{ll} \text{minimize} & \delta \\ \text{subject to} & P > 0, \quad x(0)^T P x(0) \le 1, \quad \lambda \ge 0, \quad (6.46), \\ & \begin{bmatrix} A^T P + PA + \lambda C_q^T C_q & PB_p \\ B_p^T P & -\lambda I \end{bmatrix} \le 0 \end{array}$$

Remark: As variation of this problem, we can impose a decay rate constraint on the output, that is, given $\alpha > 0$, we can compute an upper bound on the smallest γ such that the output z satisfies

$$\|z(t)\| \le \gamma e^{-\alpha t},$$

for all $t \ge 0$. This also reduces to an EVP.

6.3 Input-to-Output Properties

We finally consider the input-output behavior of the LDI

$$\begin{aligned} \dot{x} &= A(t)x + B_w(t)w, \quad x(0) = x_0, \\ z &= C_z(t)x + D_{zw}(t)w, \end{aligned} \tag{6.48}$$

where

$$\begin{bmatrix} A(t) & B_w(t) \\ C_z(t) & D_{zw}(t) \end{bmatrix} \in \Omega. \tag{6.49}$$

6.3.1 Hankel norm bounds

In this section, we assume that $D_{zw}(t) = 0$ for all t, and consider the quantity

$$\phi \triangleq \max \left\{ \int_T^\infty z^T z \, dt \;\middle|\; \begin{array}{l} \int_0^T w^T w \, dt \le 1, \quad x(0) = 0, \\ w(t) = 0, \quad t > T \ge 0 \end{array} \right\},$$

and the maximum is taken over $A(t)$, $B(t)$ and $C(t)$ satisfying (6.49). For an LTI system, $\sqrt{\phi}$ equals the *Hankel norm*, a name that we extend for convenience to LDIs.

An upper bound for ϕ can be computed by combining the ellipsoidal bounds on reachable sets from §6.1.1 and the bounds on the output energy from §6.2.1. Suppose that $P > 0$ and $Q > 0$ satisfy

$$\frac{d}{dt}\left(x^T P x\right) \le w^T w, \qquad \frac{d}{dt}\left(x^T Q x\right) \le -z^T z \tag{6.50}$$

for all x, w and z satisfying (6.48). Then, from the arguments in §6.1.1 and §6.2.1, we conclude that

$$\begin{aligned} \phi &\le \sup\left\{ \int_T^\infty z^T z \, dt \;\middle|\; x(T)^T P x(T) \le 1 \right\} \\ &\le \sup\left\{ x(T)^T Q x(T) \;\middle|\; x(T)^T P x(T) \le 1 \right\} \\ &= \lambda_{\max}(P^{-1/2} Q P^{-1/2}). \end{aligned}$$

• LTI SYSTEMS: We compute the smallest upper bound on the Hankel norm provable via quadratic functions satisfying (6.50) by computing the square root of the minimum γ such that there exist $P > 0$ and $Q > 0$ satisfying

$$\begin{aligned} &A^T P + PA + PB_w B_w^T P \le 0, \\ &A^T Q + QA + C_z^T C_z \le 0, \\ &\gamma P - Q \ge 0. \end{aligned}$$

This is a GEVP in the variables γ, P, and Q. It is possible to transform it into an EVP by introducing the new variable $\tilde{Q} = Q/\gamma$. The corresponding EVP in the variables γ, P, and $\tilde{Q}$ is then to minimize γ subject to

$$\begin{array}{l} A^T P + PA + PB_w B_w^T P \le 0, \\ A^T \tilde{Q} + \tilde{Q} A + C_z^T C_z/\gamma \le 0, \\ P - \tilde{Q} \ge 0 \end{array}$$

In this case, the optimal value is exactly the Hankel norm. It can be found analytically as $\lambda_{\max}(W_c W_o)$ where W_c and W_o are the controllability and observability Gramians, i.e., the solutions of the Lyapunov equations (6.6) and (6.38), respectively.

- POLYTOPIC LDIs: We compute the smallest upper bound on the Hankel norm provable via quadratic functions satisfying (6.50) by computing the square root of the minimum γ such that there exist $P > 0$ and $Q > 0$ satisfying

$$\begin{array}{l} A_i^T P + PA_i + PB_{w,i} B_{w,i}^T P \le 0, \\ A_i^T Q + QA_i + C_{z,i}^T C_{z,i} \le 0, \qquad i = 1, \ldots, L \\ \gamma P - Q \ge 0. \end{array}$$

This is a GEVP in the variables γ, P, and Q. Defining $\tilde{Q} = Q/\gamma$, an equivalent EVP in the variables γ, P, and $\tilde{Q}$ is to minimize γ subject to

$$\begin{array}{l} A_i^T P + PA_i + PB_{w,i} B_{w,i}^T P \le 0, \\ A_i^T \tilde{Q} + \tilde{Q} A_i + C_{z,i}^T C_{z,i}/\gamma \le 0, \qquad i = 1, \ldots, L \\ P - \tilde{Q} \ge 0 \end{array}$$

- NORM-BOUND LDIs: Similarly, the smallest upper bound on the Hankel norm provable via quadratic functions satisfying (6.50) is computed by taking the square root of the minimum γ such that there exist $P > 0$, $Q > 0$, $\lambda \ge 0$, and $\mu \ge 0$ satisfying

$$\begin{array}{l} \begin{bmatrix} A^T P + PA + \lambda C_q^T C_q & PB_p & PB_w \\ B_p^T P & -\lambda I & 0 \\ B_w^T P & 0 & -I \end{bmatrix} \le 0, \\ \begin{bmatrix} A^T Q + QA + C_z^T C_z + \mu C_q^T C_q & QB_p \\ B_p^T Q & -\mu I \end{bmatrix} \le 0, \\ \gamma P - Q \ge 0. \end{array}$$

This is a GEVP in the variables γ, P, Q, λ, and μ. If we introduce the new variables $\tilde{Q} = Q/\gamma$, $\nu = \mu/\gamma$ and equivalent EVP in the variables γ, P, $\tilde{Q}$, λ, and ν is to minimize γ such that

$$\begin{array}{l} \begin{bmatrix} A^T P + PA + \lambda C_q^T C_q & PB_p & PB_w \\ B_p^T P & -\lambda I & 0 \\ B_w^T P & 0 & -I \end{bmatrix} \le 0, \\ \begin{bmatrix} A^T \tilde{Q} + \tilde{Q} A + C_z^T C_z/\gamma + \nu C_q^T C_q & \tilde{Q} B_p \\ B_p^T \tilde{Q} & -\nu I \end{bmatrix} \le 0, \\ P - \tilde{Q} \ge 0. \end{array}$$

6.3.2 $\mathbf{L}_2$ and RMS gains

We assume $D_{zw}(t) = 0$ for simplicity of exposition. We define the $\mathbf{L}_2$ gain of the LDI (6.48) as the quantity

$$\sup_{\|w\|_2 \neq 0} \frac{\|z\|_2}{\|w\|_2}$$

where the $\mathbf{L}_2$ norm of u is $\|u\|_2^2 = \int_0^\infty u^T u \, dt$, and the supremum is taken over all nonzero trajectories of the LDI, starting from $x(0) = 0$. The LDI is said to be *nonexpansive* if its $\mathbf{L}_2$ gain is less than one.

Now, suppose there exists a quadratic function $V(\xi) = \xi^T P \xi$, $P > 0$, and $\gamma \geq 0$ such that for all t,

$$\frac{d}{dt} V(x) + z^T z - \gamma^2 w^T w \leq 0 \qquad \text{for all } x \text{ and } w \text{ satisfying (6.48).} \tag{6.51}$$

Then the $\mathbf{L}_2$ gain of the LDI is less than γ. To show this, we integrate (6.51) from 0 to T, with the initial condition $x(0) = 0$, to get

$$V(x(T)) + \int_0^T \left(z^T z - \gamma^2 w^T w \right) \, dt \leq 0.$$

since $V(x(T)) \geq 0$, this implies

$$\frac{\|z\|_2}{\|w\|_2} \leq \gamma.$$

Remark: It can be easily checked that if (6.51) holds for $V(\xi) = \xi^T P \xi$, $P > 0$, then γ is also an upper bound on the RMS gain of the LDI, where the root-mean-square (RMS) value of ξ is defined as

$$\mathbf{RMS}(\xi) \triangleq \left(\limsup_{T \to \infty} \frac{1}{T} \int_0^T \xi^T \xi \, dt \right)^{1/2},$$

and the RMS gain is defined as

$$\sup_{\mathbf{RMS}(w) \neq 0} \frac{\mathbf{RMS}(z)}{\mathbf{RMS}(w)}.$$

Now reconsider condition (6.51) for LDIs.

- LTI SYSTEMS: Condition (6.51) is equivalent to

$$\begin{bmatrix} A^T P + PA + C_z^T C_z & PB_w \\ B_w^T P & -\gamma^2 I \end{bmatrix} \leq 0. \tag{6.52}$$

Therefore, we compute the smallest upper bound on the $\mathbf{L}_2$ gain of the LTI system provable via quadratic functions by minimizing γ over the variables P and γ satisfying conditions $P > 0$ and (6.52). This is an EVP.

Remark: Assuming that (A, B_w, C_z) is minimal, this EVP gives the exact value of the $\mathbf{L}_2$ gain of the LTI system, which also equals the $\mathbf{H}_\infty$ norm of its transfer matrix, $\|C_z(sI - A)^{-1} B_w\|_\infty$.

Assume $B_w \neq 0$ and $C_z \neq 0$; then the existence of $P > 0$ satisfying (6.52) is equivalent to the existence of $Q > 0$ satisfying

$$\begin{bmatrix} AQ + QA^T + B_w B_w^T/\gamma^2 & QC_z^T \\ C_z Q & -I \end{bmatrix} \leq 0.$$

- POLYTOPIC LDIs: Condition (6.51) is equivalent to

$$\begin{bmatrix} A_i^T P + PA_i + C_{z,i}^T C_{z,i} & PB_{w,i} \\ B_{w,i}^T P & -\gamma^2 I \end{bmatrix} \leq 0, \qquad i = 1, \ldots, L. \tag{6.53}$$

Assume there exists i_0 for which $B_{w,i_0} \neq 0$, and j_0 for which $C_{z,j_0} \neq 0$. Then there exists $P > 0$ satisfying (6.53) if and only if there exists $Q > 0$ satisfying

$$\begin{bmatrix} QA_i^T + A_i Q + B_{w,i} B_{w,i}^T & QC_{z,i}^T \\ C_{z,i} Q & -\gamma^2 I \end{bmatrix} \leq 0, \qquad i = 1, \ldots, L.$$

We get the smallest upper bound on the $\mathbf{L}_2$ gain provable via quadratic functions by minimizing γ (over γ and P) subject to (6.53) and $P > 0$, which is an EVP.

- NORM-BOUND LDIs: For NLDIs, condition (6.51) is equivalent to

$$\xi^T (A^T P + PA + C_z^T C_z)\xi + 2\xi^T P (B_p \pi + B_w w) - \gamma^2 w^T w \leq 0$$

for all ξ and π satisfying

$$\pi^T \pi \leq \xi^T C_q^T C_q \xi.$$

This is true if and only if there exists $\lambda \geq 0$ such that

$$\begin{bmatrix} A^T P + PA + C_z^T C_z + \lambda C_q^T C_q & PB_p & PB_w \\ B_p^T P & -\lambda I & 0 \\ B_w^T P & 0 & -\gamma^2 I \end{bmatrix} \leq 0. \tag{6.54}$$

Therefore, we obtain the best upper bound on the $\mathbf{L}_2$ gain provable via quadratic functions by minimizing γ over the variables γ, P and λ, subject to $P > 0$, $\lambda \geq 0$ and (6.54).

If $B_w \neq 0$ and $C_z \neq 0$, then the existence of $P > 0$ and $\lambda \geq 0$ satisfying (6.54) is equivalent to the existence of $Q > 0$ and $\mu \geq 0$ satisfying

$$\begin{bmatrix} QA^T + AQ + B_w B_w^T + \mu B_p B_p^T & QC_q^T & QC_z^T \\ C_q Q & -\mu I & 0 \\ C_z Q & 0 & -\gamma^2 I \end{bmatrix} \leq 0.$$

- DIAGONAL NORM-BOUND LDIs: For DNLDIs, condition (6.51) is equivalent to

$$\xi^T (A^T P + PA + C_z^T C_z)\xi + 2\xi^T P (B_p \pi + B_w w) - \gamma^2 w^T w \leq 0$$

for all ξ and π satisfying

$$\pi_i^T \pi_i \leq (C_{q,i}\xi_i)^T (C_{q,i}\xi_i), \qquad i = 1, \ldots, n_q.$$

Using the $\mathcal{S}$-procedure, a sufficient condition is that there exist a diagonal $\Lambda \geq 0$ such that

$$\begin{bmatrix} A^TP + PA + C_z^TC_z + C_q^T\Lambda C_q & PB_p & PB_w \\ B_p^TP & -\Lambda & 0 \\ B_w^TP & 0 & -\gamma^2 I \end{bmatrix} \leq 0. \tag{6.55}$$

We obtain the best such upper bound by minimizing γ over the variables γ, P and λ, subject to $P > 0$, $\lambda \geq 0$ and (6.55).

If $B_w \neq 0$ and $C_z \neq 0$, then the existence of $P > 0$ and $\lambda \geq 0$ satisfying (6.54) is equivalent to the existence of $Q > 0$ and $M = \mathbf{diag}(\mu_1, \ldots, \mu_{n_q}) \geq 0$ satisfying

$$\begin{bmatrix} QA^T + AQ + B_wB_w^T + B_pMB_p^T & QC_q^T & QC_z^T \\ C_qQ & -M & 0 \\ C_zQ & 0 & -\gamma^2 I \end{bmatrix} \leq 0.$$

6.3.3 Dissipativity

The LDI (6.48) is said to be *passive* if every solution x with $x(0) = 0$ satisfies

$$\int_0^T w^T z \, dt \geq 0$$

for all $T \geq 0$. It is said to have dissipation η if

$$\int_0^T \left(w^T z - \eta w^T w\right) \, dt \geq 0 \tag{6.56}$$

holds for all trajectories with $x(0) = 0$ and all $T \geq 0$. Thus passivity corresponds to nonnegative dissipation. The largest dissipation of the system, i.e., the largest number η such that (6.56) holds, will be called its *dissipativity*.

Suppose that there is a quadratic function $V(\xi) = \xi^T P \xi$, $P > 0$, such that

$$\text{for all } x \text{ and } w \text{ satisfying (6.48),} \quad \frac{d}{dt}V(x) - 2w^T z + 2\eta w^T w \leq 0. \tag{6.57}$$

Then, integrating (6.57) from 0 to T with initial condition $x(0) = 0$ yields

$$V(x(T)) - \int_0^T \left(2w^T z - 2\eta w^T w\right) \, dt \leq 0.$$

Since $V(x(T)) \geq 0$, we conclude

$$\int_0^T \left(w^T z - \eta w^T w\right) \, dt \geq 0,$$

which implies that the dissipativity of the LDI is at least η. Now reconsider condition (6.57) for the various LDIs.

- LTI systems: For LTI systems, condition (6.57) is equivalent to

$$\begin{bmatrix} A^TP + PA & PB_w - C_z^T \\ B_w^TP - C_z & 2\eta I - (D_{zw}^T + D_{zw}) \end{bmatrix} \leq 0. \tag{6.58}$$

We find the largest dissipation that can be guaranteed using quadratic functions by maximizing η over the variables P and η, subject to $P > 0$ and (6.58), an EVP.

Assuming that the system (A, B_w, C_z) is minimal, the optimal value of this EVP is exactly equal to the dissipativity of the system, which can be expressed in terms of the transfer matrix $H(s) = C_z(sI - A)^{-1}B_w + D_{zw}$ as

$$\inf_{\mathbf{Re}s>0} \frac{\lambda_{\min}\left(H(s) + H(s)^*\right)}{2}. \tag{6.59}$$

This follows from the PR lemma.

Defining $Q = P^{-1}$, condition (6.58) is equivalent to

$$\begin{bmatrix} QA^T + AQ & B_w - QC_z^T \\ B_w^T - C_zQ & 2\eta I - (D_{zw}^T + D_{zw}) \end{bmatrix} \leq 0. \tag{6.60}$$

- POLYTOPIC LDIS: For PLDIs, condition (6.57) is equivalent to

$$\begin{bmatrix} A_i^TP + PA_i & PB_{w,i} - C_{z,i}^T \\ B_{w,i}^TP - C_{z,i} & 2\eta I - (D_{zw,i}^T + D_{zw,i}) \end{bmatrix} \leq 0, \quad i = 1, \ldots, L. \tag{6.61}$$

We find the largest dissipation that can be guaranteed via quadratic functions by maximizing η over the variables P and η satisfying $P > 0$ and (6.61). This is an EVP; its optimal value is a lower bound on the dissipativity of the PLDI.

If we define $Q = P^{-1}$, condition (6.57) is equivalent to

$$\begin{bmatrix} QA_i^T + A_iQ & B_{w,i} - QC_{z,i}^T \\ B_{w,i}^T - C_{z,i}Q & 2\eta I - (D_{zw,i}^T + D_{zw,i}) \end{bmatrix} \leq 0, \quad i = 1, \ldots, L. \tag{6.62}$$

- NORM-BOUND LDIS: For NLDIs, using the $\mathcal{S}$-procedure, the condition (6.57) is true if and only if there exists a nonnegative scalar λ such that

$$\begin{bmatrix} A^TP + PA + \lambda C_q^TC_q & PB_w - C_z^T & PB_p \\ B_w^TP - C_z & 2\eta I - (D_{zw}^T + D_{zw}) & 0 \\ B_p^TP & 0 & -\lambda I \end{bmatrix} \leq 0. \tag{6.63}$$

Therefore we can find the largest dissipation provable with a quadratic Lyapunov function by maximizing η (over η and P), subject to $P > 0$ and (6.63). This is an EVP. Defining $Q = P^{-1}$, condition (6.57) is equivalent to the existence of $\mu \geq 0$ satisfying

$$\begin{bmatrix} QA^T + AQ + \mu B_pB_p^T & B_w - QC_z^T & QC_q^T \\ B_w^T - C_zQ & 2\eta I - (D_{zw}^T + D_{zw}) & 0 \\ C_qQ & 0 & -\mu I \end{bmatrix} \leq 0. \tag{6.64}$$

6.3.4 Diagonal scaling for componentwise results

We assume for simplicity that $D_{zw}(t) = 0$. Assuming that the system (6.48) has as many inputs as outputs, we consider the system

$$\begin{aligned} \dot{x} &= A(t)x + B_w(t)T^{-1}\tilde{w}, \qquad x(0) = 0, \\ \tilde{z} &= TC_z(t)x. \end{aligned} \tag{6.65}$$

where T is a positive-definite diagonal matrix, which has the interpretation of a *scaling*. We will see that scaling enables us to conclude "componentwise" results for the system (6.65).

Consider for example, the "scaled $\mathbf{L}_2$ gain"

$$\alpha = \inf_{\substack{T \text{ diagonal}, \\ T>0}} \sup_{\|\tilde{w}\|_2 \neq 0} \frac{\|\tilde{z}\|_2}{\|\tilde{w}\|_2}.$$

The scaled $\mathbf{L}_2$ gain has the following interpretation:

$$\max_{i=1,\ldots,n_z} \sup_{\|w_i\|_2 \neq 0} \frac{\|z_i\|}{\|w_i\|} \leq \alpha.$$

We show this as follows: For every fixed scaling $T = \mathbf{diag}(t_1, \ldots, t_{n_z})$,

$$\begin{array}{rcl} \sup_{\|\tilde{w}\|_2 \neq 0} \dfrac{\|\tilde{z}\|_2^2}{\|\tilde{w}\|_2^2} & = & \sup_{\|w\|_2 \neq 0} \dfrac{\sum_i^{n_z} t_i^2 \|z_i\|_2^2}{\sum_i^{n_z} t_i^2 \|w_i\|_2^2} \\ & \geq & \sup_{\|w_i\|_2 \neq 0} \dfrac{\|z_i\|_2^2}{\|w_i\|_2^2}. \end{array}$$

Since the inequality is true for all diagonal nonsingular T, the desired inequality follows. We now show how we can obtain bounds on the scaled $\mathbf{L}_2$ gain using LMIs for LTI systems and LDIs.

• LTI systems: The $\mathbf{L}_2$ gain for the LTI system scaled by T is guaranteed to be less than γ if there exists $P > 0$ such that the LMI

$$\left[\begin{array}{cc} A^T P + PA + C_z^T S C_z & PB_w \\ B_w^T P & -\gamma^2 S \end{array} \right] < 0$$

holds with $S = T^T T$. The smallest scaled $\mathbf{L}_2$ gain of the system can therefore be computed as a GEVP.

• Polytopic LDIs: For PLDIs, the scaled $\mathbf{L}_2$ gain is guaranteed to be lower than γ if there exists $S > 0$, with S diagonal, and $P > 0$ which satisfy

$$\left[\begin{array}{cc} A_i^T P + PA_i + C_{z,i}^T S C_{z,i} & PB_{w,i} \\ B_{w,i}^T P & -\gamma^2 S \end{array} \right] < 0, \quad i = 1, \ldots, L. \tag{6.66}$$

The optimal scaled $\mathbf{L}_2$ gain γ is therefore obtained by minimizing γ over (γ, P and S) subject to (6.66), $P > 0$, $S > 0$ and S diagonal. This is a GEVP.

• Norm-bound LDIs: For NLDIs, the scaled $\mathbf{L}_2$ gain (with scaling T) is less than γ if there exist $\lambda \geq 0$, $P > 0$, and $S > 0$, S diagonal, such that the following LMI holds:

$$\left[\begin{array}{ccc} A^T P + PA + C_z^T S C_z + \lambda C_q^T C_q & PB_p & PB_w \\ B_p^T P & -\lambda I & 0 \\ B_w^T P & 0 & -\gamma^2 S \end{array} \right] \leq 0. \tag{6.67}$$

Therefore, minimizing γ^2 subject to the constraint (6.67) is an EVP.

Notes and References

Integral quadratic constraints

In §8.2 (and §10.9) we consider a generalization of the NLDI in which the pointwise condition $p^T p \le q^T q$ is replaced by the integral constraint

$$\int_0^\infty p^T p \, dt \le \int_0^\infty q^T q \, dt.$$

Many of the results derived for NLDIs and DNLDIs then become necessary and sufficient. Integral quadratic constraints were introduced by Yakubovich. For a general study of these, we refer the reader to the articles of Yakubovich (see [YAK92] and references therein), and also Megretsky [MEG93, MEG92A, MEG92B].

Reachable sets for componentwise unit-energy inputs

We study in greater detail the set of states reachable with componentwise unit-energy inputs for LTI systems (see §6.1.2). We begin by observing that a point x_0 belongs to the reachable set if and only if the optimal value of the problem

$$\begin{array}{ll} \text{minimize} & \max_i \int_0^\infty w_i^2 \, dt \\ \text{subject to} & \dot{x} = -Ax + B_w w, \quad x(0) = x_0, \qquad \lim_{t\to\infty} x(t) = 0 \end{array}$$

is less than one. This is a multi-criterion convex quadratic problem, considered in §10.8. In this case, the problem reduces to checking whether the LMI

$$\begin{array}{l} x_0^T P x_0 < 1, \qquad \begin{bmatrix} A^T P + PA & PB \\ B^T P & -R \end{bmatrix} \le 0, \\ P > 0, \qquad R > 0, \text{ diagonal}, \qquad \mathbf{Tr}\, R = 1 \end{array}$$

is feasible. This shows that the reachable set is the intersection of ellipsoids $\left\{\xi \mid \xi^T P \xi \le 1\right\}$ satisfying equation (6.14).

A variation of this problem is considered in [SZ92], where the authors consider inputs w satisfying $\int_0^\infty w w^T \, dt \le Q$, or equivalently

$$\int_0^\infty w^T W w \, dt \le \mathbf{Tr}\, WQ \quad \text{for every } W \ge 0.$$

They get necessary conditions characterizing the corresponding reachable set. We now derive LMI conditions that are both necessary and sufficient.

As before, x_0 belongs to the reachable set if and only the optimal value of the problem

$$\begin{array}{ll} \text{minimize} & \int_0^\infty w^T W w \, dt \\ \text{subject to} & \dot{x} = -Ax + B_w w, \quad x(0) = x_0, \qquad \lim_{t\to\infty} x(t) = 0 \end{array}$$

is less than $\mathbf{Tr}\, WQ$ for every $W \ge 0$. This is equivalent to infeasibility of the LMI (in the variables P and W):

$$x_0^T P x_0 - \mathbf{Tr}\, WQ \ge 0, \qquad P > 0, \qquad \begin{bmatrix} A^T P + PA & PB \\ B^T P & -W \end{bmatrix} \le 0.$$

These are precisely the conditions obtained in [SZ92].

Reachable sets with unit-peak inputs

Schweppe [SCH73, §4.3.3] considers the more general problem of *time-varying* ellipsoidal approximations of reachable sets with unit-peak inputs for time-varying systems; the positive-definite matrix describing the ellipsoidal approximation satisfies a matrix differential equation, terms of which closely resemble those in the matrix inequalities described in this book. Sabin and Summers [SS90] study the approximation of the reachable set via quadratic functions. A survey of the available techniques can be found in the article by Gayek [GAY91]. See also [FG88].

The technique used to derive LMI (6.23) can be interpreted as an exponential time-weighting procedure. Fix $\alpha > 0$. Then for every $T > 0$, rewrite LDI (6.1), with new exponentially time-weighted variables $x_T(t) = e^{\alpha(t-T)/2}x(t)$ and $v(t) = e^{\alpha(t-T)/2}w(t)$ as

$$\dot{x}_T = \left(A(t) + \frac{\alpha}{2}I\right)x_T + B_w(t)v, \quad x_T(0) = 0.$$

Since $w(t)^T w(t) \le 1$, we have $\int_0^T v^T v \, d\tau \le 1/\alpha$.

Now suppose that $P > 0$ satisfies condition (6.23), which we rewrite for convenience as

$$\begin{bmatrix} (A + \alpha I/2)^T P + P(A + \alpha I/2) & PB_w \\ B_w^T P & -\alpha I \end{bmatrix} \le 0. \tag{6.68}$$

The results of §6.1.1 imply that $x_T(T)$ satisfies $x_T(T)^T P x_T(T) \le 1$; therefore, $x(T)^T P x(T) \le 1$. Since LMI (6.68) is independent of T, the ellipsoid $\left\{x \mid x^T P x \le 1\right\}$ contains the reachable set with unit-peak inputs.

Bounds on overshoot

The results of §6.1.3 can be used to find a bound on the step response peak for LDIs. Consider a single-input single-output LDI, subject to a unit-step input:

$$\begin{aligned} \dot{x} &= A(t)x + b_w(t), \quad x(0) = 0, \\ s &= c_z(t)x, \end{aligned} \qquad \begin{bmatrix} A(t) & b_w(t) \\ c_z(t) & 0 \end{bmatrix} \in \Omega.$$

Since a unit-step is also a unit-peak input, an upper bound on $\max_{t\ge 0} |s(t)|$ (i.e., the *overshoot*) can be found by combining the results on reachable sets with unit-peak inputs (§6.1.3) and bounds on output peak (§6.2.2).

Suppose there exist $Q > 0$ and $\alpha > 0$ satisfying (6.25),

$$A(t)Q + QA(t)^T + \alpha Q + \frac{1}{\alpha} b_w(t) b_w(t)^T < 0.$$

Then, from §6.2.2, we conclude that $\max_{t\ge 0} |s(t)|$ does not exceed M, where $M^2 = \max_{t\ge 0} c_z(t)^T Q c_z(t)$. This can be used to find LMI-based bounds on overshoot for LTI systems, PLDIs and NLDIs. We note that the resulting bounds can be quite conservative. These LMIs can be used to determine state-feedback matrices as well (see the Notes and References of Chapter 7).

Bounds on impulse response peak for LTI systems

The impulse response $h(t) = ce^{At}b$ of the single-input single-output LTI system

$$\dot{x} = Ax + bu, \quad x(0) = 0, \qquad y = cx$$

is just the output y with initial condition $x(0) = b$ and zero input. Therefore, we can compute a bound on the impulse response peak by combining the results on invariant ellipsoids from §5.2 with those on the peak value of the output from §6.2.2. The bound obtained this way can be quite conservative. For instance, consider the LTI system with transfer function

$$H(s) = \frac{1}{s + s_1} \prod_{i=2}^{n} \frac{s - s_i}{s + s_i},$$

where $s_{i+1} \gg s_i > 0$, i.e., the dynamics of this system are widely spaced. The bound on the peak value of the impulse response of the system computed via the EVP (6.47) turns out to be $2n - 1$ times the actual maximum value of the impulse response, where n is the dimension of a minimal realization of the system [FER93]. We conjecture that the bound can be no more conservative, that is, for every LTI system, the bound on the peak of the impulse response computed via the EVP (6.47) can be no more than $2n - 1$ times the actual maximum value of the impulse response.

Examples for which the bound obtained via the EVP (6.47) is sharp include the case of passive LTI systems. In this case, we know there exists a positive-definite P such that $A^T P + PA \leq 0$, $Pb = c^T$, so that the maximum value of the impulse response is cb and is attained for $t = 0$. This is an illustration of the fact that passive systems have nice "peaking" properties, which is used in nonlinear control [KOK92].

Chapter 7

State-Feedback Synthesis for LDIs

7.1 Static State-Feedback Controllers

We consider the LDI

$$\dot{x} = A(t)x + B_w(t)w + B_u(t)u, \qquad z = C_z(t)x + D_{zw}(t)w + D_{zu}(t)u,$$
$$\begin{bmatrix} A(t) & B_w(t) & B_u(t) \\ C_z(t) & D_{zw}(t) & D_{zu}(t) \end{bmatrix} \in \Omega, \tag{7.1}$$

where Ω has one of our special forms (i.e., singleton, polytope, image of a unit ball under a matrix linear-fractional mapping). Here $u : \mathbf{R}_+ \to \mathbf{R}^{n_u}$ is the *control input* and $w : \mathbf{R}_+ \to \mathbf{R}^{n_w}$ is the *exogenous input signal.*

Let $K \in \mathbf{R}^{n_u \times n}$, and suppose that $u = Kx$. Since the control input is a linear function of the state, this is called (linear, constant) *state-feedback*, and the matrix K is called the *state-feedback gain.* This yields the *closed-loop* LDI

$$\begin{aligned} \dot{x} &= (A(t) + B_u(t)K)\,x + B_w(t)w, \\ z &= (C_z(t) + D_{zu}(t)K)\,x + D_{zw}(t)w. \end{aligned} \tag{7.2}$$

In this chapter we consider the state-feedback synthesis problem, i.e., the problem of finding a matrix K so that the closed-loop LDI (7.2) satisfies certain properties or specifications, e.g., stability.

> **Remark**: Using the idea of global linearization described in §4.3, the methods of this chapter can be used to synthesize a linear state-feedback for some nonlinear, time-varying systems. As an example, we can synthesize a state-feedback for the system
>
> $$\dot{x} = f(x, w, u, t), \qquad z = g(x, w, u, t),$$
>
> provided we have
>
> $$\begin{bmatrix} \dfrac{\partial f}{\partial x} & \dfrac{\partial f}{\partial w} & \dfrac{\partial f}{\partial u} \\ \dfrac{\partial g}{\partial x} & \dfrac{\partial g}{\partial w} & \dfrac{\partial g}{\partial u} \end{bmatrix} (x, w, u, t) \in \Omega,$$
>
> and $f(0,0,0,t) = 0$, $g(0,0,0,t) = 0$, for all x, t, w and u.

7.2 State Properties

We consider the LDI

$$\dot{x} = A(t)x + B_u(t)u, \qquad [A(t)\ B_u(t)] \in \Omega. \tag{7.3}$$

• LTI SYSTEMS: For LTI systems, (7.3) becomes

$$\dot{x} = Ax + B_u u. \tag{7.4}$$

• POLYTOPIC LDIS: PLDIs are given by

$$\dot{x} = A(t)x + B_u(t)u, \qquad [A(t)\ \ B_u(t)] \in \mathbf{Co}\left\{[A_1\ \ B_{u,1}], \ldots. [A_L\ \ B_{u,L}]\right\}. \tag{7.5}$$

• NORM-BOUND LDIS: For NLDIs, (7.3) becomes

$$\begin{array}{ll} \dot{x} = Ax + B_u u + B_p p, & q = C_q x + D_{qu} u + D_{qp} p \\ p = \Delta(t) q, & \|\Delta(t)\| \le 1. \end{array} \tag{7.6}$$

Equivalently, we have

$$\dot{x} = Ax + B_u u + B_p p, \qquad q = C_q x + D_{qp} p + D_{qu} u, \qquad p^T p \le q^T q.$$

• DIAGONAL NORM-BOUND LDIS: For DNLDIs, equation (7.3) becomes

$$\begin{array}{ll} \dot{x} = Ax + B_u u + B_p p, & q = C_q x + D_{qp} p + D_{qu} u, \\ p_i = \delta_i(t) q_i, & |\delta_i(t)| \le 1, \quad i = 1, \ldots, L. \end{array} \tag{7.7}$$

Equivalently, we have

$$\begin{array}{ll} \dot{x} = Ax + B_u u + B_p p, & q = C_q x + D_{qp} p + D_{qu} u, \\ |p_i| \le |q_i|, \quad i = 1, \ldots, n_q. & \end{array}$$

7.2.1 Quadratic stabilizability

The system (7.1) is said to be *quadratically stabilizable* (via linear state-feedback) if there exists a state-feedback gain K such that the closed-loop system (7.2) is quadratically stable (hence, stable). Quadratic stabilizability can be expressed as an LMIP.

• LTI SYSTEMS: Let us fix the matrix K. The LTI system (7.4) is (quadratically) stable if and only if there exists $P > 0$ such that

$$(A + B_u K)^T P + P(A + B_u K) < 0,$$

or equivalently, there exists $Q > 0$ such that

$$Q(A + B_u K)^T + (A + B_u K)Q < 0. \tag{7.8}$$

Neither of these conditions is jointly convex in K and P or Q, but by a simple change of variables we can obtain an equivalent condition that is an LMI.

Define $Y = KQ$, so that for $Q > 0$ we have $K = YQ^{-1}$. Substituting into (7.8) yields

$$AQ + QA^T + B_u Y + Y^T B_u^T < 0, \tag{7.9}$$

which is an LMI in Q and Y. Thus, the system (7.4) is quadratically stabilizable if and only if there exist $Q > 0$ and Y such that the LMI (7.9) holds. If this LMI is feasible, then the quadratic function $V(\xi) = \xi^T Q^{-1} \xi$ proves (quadratic) stability of system (7.4) with state-feedback $u = YQ^{-1}x$.

Remark: We will use the simple change of variables $Y = KQ$ many times in the sequel. It allows us to recast the problem of finding a state-feedback gain as an LMIP with variables that include Q and Y; the state-feedback gain K is recovered as $K = YQ^{-1}$

An alternate equivalent condition for (quadratic) stabilizability, involving fewer variables, can be derived using the elimination of matrix variables described in §2.6.2: There exist $Q > 0$ and a scalar σ such that

$$AQ + QA^T - \sigma B_u B_u^T < 0. \tag{7.10}$$

Since we can always assume $\sigma > 0$ in LMI (7.10), and since the LMI is homogeneous in Q and σ, we can without loss of generality take $\sigma = 1$, thus reducing the number of variables by one. If $Q > 0$ satisfies the LMI (7.10), a stabilizing state-feedback gain is given by $K = -(\sigma/2) B_u^T Q^{-1}$.

From the elimination procedure of §2.6.2, another equivalent condition is

$$\tilde{B}_u^T \left(AQ + QA^T \right) \tilde{B}_u < 0, \tag{7.11}$$

where $\tilde{B}_u$ is an orthogonal complement of B_u. For any $Q > 0$ satisfying (7.11), a stabilizing state-feedback gain is $K = -(\sigma/2) B_u^T Q^{-1}$, where σ is any scalar such that (7.10) holds (condition (7.11) implies that such a scalar exists).

For LTI systems, these LMI conditions are necessary and sufficient for stabilizability. In terms of linear system theory, stabilizability is equivalent to the condition that every unstable mode be controllable; it is not hard to show that this is equivalent to feasibility of the LMIs (7.9), (7.11), or (7.10).

• Polytopic LDIs: For the PLDI (7.5), the same argument applies. Quadratic stabilizability is equivalent to the existence of $Q > 0$ and Y with

$$QA_i^T + A_i Q + B_{u,i} Y + Y^T B_{u,i}^T < 0, \quad i = 1, \ldots, L. \tag{7.12}$$

• Norm-bound LDIs: With $u = Kx$, the NLDI (7.6) is quadratically stable (see LMI (5.14)) if there exist $Q > 0$ and $\mu > 0$ such that

$$\begin{bmatrix} \begin{pmatrix} AQ + QA^T + B_u KQ \\ + QK^T B_u^T + \mu B_p B_p^T \end{pmatrix} & \mu B_p D_{qp}^T + Q(C_q + D_{qu}K)^T \\ \mu D_{qp} B_p^T + (C_q + D_{qu}K)Q & -\mu(I - D_{qp} D_{qp}^T) \end{bmatrix} < 0.$$

This condition has a simple frequency-domain interpretation: The $\mathbf{H}_\infty$ norm of the transfer matrix from p to q for the LTI system $\dot{x} = (A + B_u K)x + B_p p$, $q = (C_q + D_{qu}K)x + D_{qp}p$ is less than one.

With $Y = KQ$, we conclude that the NLDI is quadratically stabilizable if there exist $Q > 0$, $\mu > 0$ and Y such that

$$\begin{bmatrix} \begin{pmatrix} AQ + QA^T + \mu B_p B_p^T \\ + B_u Y + Y^T B_u^T \end{pmatrix} & \mu B_p D_{qp}^T + QC_q^T + Y^T D_{qu}^T \\ \mu D_{qp} B_p^T + C_q Q + D_{qu} Y & -\mu(I - D_{qp} D_{qp}^T) \end{bmatrix} < 0. \tag{7.13}$$

Using the elimination procedure of §2.6.2, we obtain the equivalent conditions:

$$\begin{bmatrix} \begin{pmatrix} AQ + QA^T \\ -\sigma B_u B_u^T + \mu B_p B_p^T \end{pmatrix} & QC_q^T + \mu B_p D_{qp}^T - \sigma B_u D_{qu}^T \\ C_q Q + \mu D_{qp} B_p^T - \sigma D_{qu} B_u^T & -\mu(I - D_{qp} D_{qp}^T) - \sigma D_{qu} D_{qu}^T \end{bmatrix} < 0, \tag{7.14}$$
$$-\mu(I - D_{qp}D_{qp}^T) < 0.$$

The latter condition is always satisfied since the NLDI is well-posed. By homogeneity, we can set $\mu = 1$. If $Q > 0$ and $\sigma > 0$ satisfy (7.14) with $\mu = 1$, a stabilizing state-feedback gain is $K = (-\sigma/2)B_u^T Q^{-1}$.

An alternate, equivalent condition for quadratic stabilizability is expressed in terms of the orthogonal complement of $[B_u^T \; D_{qu}^T]^T$, which we denote by $\tilde{G}$:

$$\tilde{G}^T \begin{bmatrix} AQ + QA^T + \mu B_p B_p^T & QC_q^T + \mu B_p D_{qp}^T \\ C_q Q + \mu D_{qp} B_p^T & -\mu(I - D_{qp} D_{qp}^T) \end{bmatrix} \tilde{G} < 0. \tag{7.15}$$

(Again, we can freely set $\mu = 1$ in this LMI.)

In the remainder of this chapter we will assume that D_{qp} in (7.6) is zero in order to simplify the discussion; all the following results can be extended to the case in which D_{qp} is nonzero.

• DIAGONAL NORM-BOUND LDIs: With $u = Kx$, the DNLDI (7.7) is quadratically stable (see LMI (5.16)) if there exist $Q > 0$, $M = \mathbf{diag}(\mu_1, \ldots, \mu_{n_q}) > 0$ satisfying

$$\begin{bmatrix} \begin{pmatrix} AQ + QA^T + B_u K Q \\ + QK^T B_u^T + B_p M B_p^T \end{pmatrix} & B_p M D_{qp}^T + Q(C_q + D_{qu}K)^T \\ D_{qp} M B_p^T + (C_q + D_{qu} K) Q & -(M - D_{qp} M D_{qp}^T) \end{bmatrix} < 0.$$

This condition leads to a simple frequency-domain interpretation for quadratic stabilizability of DNLDIs: Let H denote the transfer matrix from p to q for the LTI system $\dot{x} = (A + B_u K)x + B_p p$, $q = (C_q + D_{qu}K)x + D_{qp}p$. Then the DNLDI is quadratically stabilizable if there exists K such that for some diagonal positive-definite matrix M, $\|M^{-1/2} H M^{1/2}\|_\infty \le 1$. There is no analytical method for checking this condition.

With $Y = KQ$, we conclude that the NLDI is quadratically stabilizable if there exist $Q > 0$, $M > 0$ and diagonal, and Y such that

$$\begin{bmatrix} \begin{pmatrix} AQ + QA^T + B_p M B_p^T \\ + B_u Y + Y^T B_u^T \end{pmatrix} & B_p M D_{qp}^T + QC_q^T + Y^T D_{qu}^T \\ D_{qp} M B_p^T + C_q Q + D_{qu} Y & -(M - D_{qp} M D_{qp}^T) \end{bmatrix} < 0.$$

As before, we can eliminate the variable Y using the elimination procedure of §2.6.2, and obtain equivalent LMIs in fewer variables.

7.2.2 Holdable ellipsoids

Just as quadratic stability can also be interpreted in terms of invariant ellipsoids, we can interpret quadratic stabilizability in terms of *holdable* ellipsoids. We say that the ellipsoid

$$\mathcal{E} = \{\xi \in \mathbf{R}^n \mid \xi^T Q^{-1} \xi \le 1\}$$

is holdable for the system (7.3) if there exists a state-feedback gain K such that $\mathcal{E}$ is invariant for the system (7.3) with $u = Kx$. Therefore the LMIs described in the previous section also characterize holdable ellipsoids for the system (7.3).

With this parametrization of quadratic stabilizability and holdable ellipsoids as LMIs in $Q > 0$ and Y, we can solve various optimization problems, such as finding the coordinate transformation that minimizes the condition number of Q, imposing norm constraints on the input $u = Kx$, etc.

Remark: In Chapter 5, we saw that quadratic stability and invariant ellipsoids were characterized by LMIs in a variable $P > 0$ and *also* its inverse $Q = P^{-1}$. In contrast, quadratic stabilizability and ellipsoid holdability can be expressed as LMIs only in variables Q and Y; it is not possible to rewrite these LMIs as LMIs with Q^{-1} as a variable. This restricts the extension of some of the results from Chapter 5 (and Chapter 6) to state-feedback synthesis. As a rule of thumb, results in the analysis of LDIs that are expressed as LMIs in the variable Q can be extended to state-feedback synthesis, with a few exceptions; results expressed as LMIs in P are not. For example, we will not be able to compute the minimum volume holdable ellipsoid that contains a given polytope $\mathcal{P}$ (see problem (5.33)) as an optimization problem over LMIs; however, we will be able to compute the minimum diameter holdable ellipsoid (see problem (5.34)) containing $\mathcal{P}$.

7.2.3 Constraints on the control input

When the initial condition is known, we can find an upper bound on the norm of the control input $u(t) = Kx(t)$ as follows. Pick $Q > 0$ and Y which satisfy the quadratic stabilizability condition (either (7.9), (7.12) or (7.13)), and in addition $x(0)^T Q^{-1} x(0) \leq 1$. This implies that $x(t)$ belongs to $\mathcal{E}$ for all $t \geq 0$, and consequently,

$$\begin{aligned} \max_{t\geq 0} \|u(t)\| &= \max_{t\geq 0} \|YQ^{-1}x(t)\| \\ &\leq \max_{x\in\mathcal{E}} \|YQ^{-1}x\| \\ &= \lambda_{\max}(Q^{-1/2}Y^TYQ^{-1/2}). \end{aligned}$$

Therefore, the constraint $\|u(t)\| \leq \mu$ is enforced at all times $t \geq 0$ if the LMIs

$$\begin{bmatrix} 1 & x(0)^T \\ x(0) & Q \end{bmatrix} \geq 0, \qquad \begin{bmatrix} Q & Y^T \\ Y & \mu^2 I \end{bmatrix} \geq 0 \tag{7.16}$$

hold, where $Q > 0$ and Y satisfy the stabilizability conditions (either (7.9), (7.12) or (7.13)).

We can extend this to the case where $x(0)$ lies in an ellipsoid or polytope. For example suppose we require (7.16) to hold for all $\|x(0\| \leq 1$. This is easily shown to be equivalent to

$$Q \geq I, \qquad \begin{bmatrix} Q & Y^T \\ Y & \mu^2 I \end{bmatrix} \geq 0.$$

As another extension we can handle constraints on

$$\|u(t)\|_{\max} \stackrel{\Delta}{=} \max_i |u_i(t)|$$

in a similar manner:

$$\begin{aligned}\max_{t\geq 0}\|u(t)\|_{\max} &= \max_{t\geq 0}\|YQ^{-1}x(t)\|_{\max}\\ &\leq \max_{x\in\mathcal{E}}\|YQ^{-1}x\|_{\max}\\ &= \max_i\left(YQ^{-1}Y^T\right)_{ii}.\end{aligned}$$

Therefore, the constraint $\|u(t)\|_{\max} \leq \mu$ for $t \geq 0$ is implied by the LMI

$$\begin{bmatrix} 1 & x(0)^T \\ x(0) & Q \end{bmatrix} \geq 0, \qquad \begin{bmatrix} X & Y \\ Y^T & Q \end{bmatrix} \geq 0, \qquad X_{ii} \leq \mu^2,$$

where once again, $Q > 0$ and Y satisfy the stabilizability conditions (either (7.9), (7.12) or (7.13)).

7.3 Input-to-State Properties

We next consider the LDI

$$\dot{x} = A(t)x + B_w(t)w + B_u(t)u. \tag{7.17}$$

• LTI SYSTEMS: For LTI systems, equation (7.17) becomes $\dot{x} = Ax + B_w w + B_u u$.

• POLYTOPIC LDIS: PLDIs are given by $\dot{x} = A(t)x + B_w(t)w + B_u(t)u$, where $[A(t) \;\; B_w(t) \;\; B_u(t)] \in \mathbf{Co}\{[A_1 \;\; B_{w,1} \;\; B_{u,1}], \ldots, [A_L \;\; B_{w,L} \;\; B_{u,L}]\}$.

• NORM-BOUND LDIS: For NLDIs, equation (7.17) becomes

$$\begin{aligned}&\dot{x} = Ax + B_u u + B_p p + B_w w, \qquad q = C_q x + D_{qu} u\\ &p = \Delta(t)q, \qquad \|\Delta(t)\| \leq 1\end{aligned}$$

which we can also express as

$$\dot{x} = Ax + B_u u + B_p p + B_w w, \qquad q = C_q x + D_{qu} u, \qquad p^T p \leq q^T q.$$

• DIAGONAL NORM-BOUND LDIS: For DNLDIs, equation (7.17) becomes

$$\begin{aligned}&\dot{x} = Ax + B_u u + B_p p + B_w w, \qquad q = C_q x + D_{qu} u\\ &p_i = \delta_i(t) q_i, \qquad |\delta_i(t)| \leq 1, \quad i = 1, \ldots, L.\end{aligned}$$

Equivalently, we have

$$\begin{aligned}&\dot{x} = Ax + B_u u + B_p p + B_w w, \qquad q = C_q x + D_{qu} u,\\ &|p_i| \leq |q_i|, \quad i = 1, \ldots, n_q.\end{aligned}$$

7.3.1 Reachable sets with unit-energy inputs

For the system (7.17) with $u = Kx$, the set of states reachable with unit energy is defined as

$$\mathcal{R}_{\rm ue} \stackrel{\Delta}{=} \left\{ x(T) \;\middle|\; \begin{array}{c} x,\ w,\ u \text{ satisfy (7.17)}, \quad u = Kx, \quad x(0) = 0, \\ \displaystyle\int_0^T w^T w\,dt \leq 1, \quad T \geq 0 \end{array} \right\}.$$

We will now derive conditions under which there exists a state-feedback gain K guaranteeing that a given ellipsoid $\mathcal{E} = \{\xi \in \mathbf{R}^n \mid \xi^T Q^{-1} \xi \leq 1\}$ contains $\mathcal{R}_{\rm ue}$.

• LTI SYSTEMS: From §6.1.1, the ellipsoid $\mathcal{E}$ contains $\mathcal{R}_{\rm ue}$ for the system (7.4) for some state-feedback gain K if K satisfies

$$AQ + QA^T + B_u KQ + QK^T B_u^T + B_w B_w^T \leq 0.$$

Setting $KQ = Y$, we conclude that $\mathcal{E} \supseteq \mathcal{R}_{\rm ue}$ for some K if there exist Y such that

$$QA^T + AQ + B_u Y + Y^T B_u^T + B_w B_w^T \leq 0.$$

For any $Q > 0$ and Y satisfying this LMI, the state-feedback gain $K = YQ^{-1}$ is such that the ellipsoid $\mathcal{E}$ contains the reachable set for the closed-loop system.

Using the elimination procedure of §2.6.2, we eliminate the variable Y to obtain the LMI in $Q > 0$ and the variable σ:

$$QA^T + AQ - \sigma B_u B_u^T + B_w B_w^T \leq 0. \tag{7.18}$$

For any $Q > 0$ and σ satisfying this LMI, a corresponding state-feedback gain is given by $K = -(\sigma/2) B_u^T Q^{-1}$. Another equivalent condition for $\mathcal{E}$ to contain $\mathcal{R}_{\rm ue}$ is

$$\tilde{B}_u^T \left(AQ + QA^T + B_w B_w^T\right) \tilde{B}_u \leq 0, \tag{7.19}$$

where $\tilde{B}_u$ is an orthogonal complement of B_u.

• POLYTOPIC LDIS: For PLDIs, $\mathcal{E}$ contains $\mathcal{R}_{\rm ue}$ for some state-feedback gain K if there exist $Q > 0$ and Y such that the following LMI holds (see LMI (6.9)):

$$QA_i^T + A_i Q + B_{u,i} Y + Y^T B_{u,i}^T + B_{w,i} B_{w,i}^T < 0, \; i = 1, \ldots, L.$$

• NORM-BOUND LDIS: For NLDIs, $\mathcal{E}$ contains $\mathcal{R}_{\rm ue}$ for some state-feedback gain K if there exist $\mu > 0$ and Y such that the following LMI holds (see LMI (6.11)):

$$\begin{bmatrix} QA^T + AQ + B_u Y + Y^T B_u^T + B_w B_w^T + \mu B_p B_p^T & (C_q Q + D_{qu} Y)^T \\ C_q Q + D_{qu} Y & -\mu I \end{bmatrix} \leq 0.$$

We can eliminate Y to obtain the LMI in $Q > 0$ and σ that guarantees that $\mathcal{E}$ contains the reachable set $\mathcal{R}_{\rm ue}$ for some state-feedback gain K:

$$\begin{bmatrix} \begin{pmatrix} AQ + QA^T - \sigma B_u B_u^T \\ +\mu B_p B_p^T + B_w B_w^T \end{pmatrix} & QC_q^T + \mu B_p D_{qp}^T - \sigma B_u D_{qu}^T \\ C_q Q + \mu D_{qp} B_p^T - \sigma D_{qu} B_u^T & -\mu(I - D_{qp} D_{qp}^T) - \sigma D_{qu} D_{qu}^T \end{bmatrix} < 0. \tag{7.20}$$

The corresponding state-feedback gain is $K = (-\sigma/2) B_u^T Q^{-1}$.

An equivalent condition is expressed in terms of $\tilde{G}$, the orthogonal complement of $[B_u^T \; D_{qu}^T]^T$:

$$\tilde{G}^T \begin{bmatrix} \begin{pmatrix} AQ + QA^T \\ +\mu B_p B_p^T + B_w B_w^T \end{pmatrix} & QC_q^T + \mu B_p D_{qp}^T \\ C_q Q + \mu D_{qp} B_p^T & -\mu(I - D_{qp} D_{qp}^T) \end{bmatrix} \tilde{G} < 0.$$

• DIAGONAL NORM-BOUND LDIS: For DNLDIs, $\mathcal{E}$ contains $\mathcal{R}_{\rm ue}$ for some state-feedback gain K if there exist $Q > 0$, $M > 0$ and diagonal, and Y such that the following LMI holds (see LMI (6.12)):

$$\begin{bmatrix} QA^T + AQ + B_uY + Y^TB_u^T + B_wB_w^T + B_pMB_p^T & (C_qQ + D_{qu}Y)^T \\ C_qQ + D_{qu}Y & -M \end{bmatrix} \leq 0.$$

Using the elimination procedure of §2.6.2, we can eliminate the variable Y to obtain an equivalent LMI in fewer variables.

Using these LMIs (see §6.1.1 for details):

- We can find a state-feedback gain K such that a given point x_0 lies outside the set of reachable states for the system (7.3).
- We can find a state-feedback gain K such that the set of reachable states for the system (7.3) lies in a given half-space. This result can be extended to check if the reachable set is contained in a polytope. In this case, in contrast with the results in Chapter 6, we must use the *same* outer ellipsoidal approximation to check different faces. (This is due to the coupling induced by the new variable $Y = KQ$.)

7.3.2 Reachable sets with componentwise unit-energy inputs

With $u = Kx$, the set of reachable states for inputs with componentwise unit-energy for the system (7.17) is defined as

$$\mathcal{R}_{\rm uce} \triangleq \left\{ x(T) \;\middle|\; \begin{array}{c} x,\ w,\ u \text{ satisfy } (7.17), \quad u = Kx, \quad x(0) = 0, \\ \displaystyle\int_0^T w_i^T w_i \, dt \leq 1, \qquad T \geq 0 \end{array} \right\}.$$

We now consider the existence of the state-feedback gain K such that the ellipsoid $\mathcal{E} = \{\xi \in \mathbf{R}^n \mid \xi^T Q^{-1} \xi \leq 1\}$ contains $\mathcal{R}_{\rm uce}$ for LTI systems and LDIs.

• LTI SYSTEMS: From LMI (6.15), $\mathcal{E} \supseteq \mathcal{R}_{\rm uce}$ for the LTI system (7.4) if there exists diagonal $R > 0$ with unit trace such that the LMI

$$\begin{bmatrix} QA^T + AQ + B_uY + Y^TB_u^T & B_w \\ B_w^T & -R \end{bmatrix} \leq 0.$$

• POLYTOPIC LDIS: For PLDIs, $\mathcal{E} \supseteq \mathcal{R}_{\rm uce}$ for some state-feedback gain K if there exist $Q > 0$, $Y (= KQ)$, and a diagonal $R > 0$ with unit trace (see LMI (6.17)), such that

$$\begin{bmatrix} QA_i^T + A_iQ + B_{u,i}Y + Y^TB_{u,i}^T & B_{w,i} \\ B_{w,i}^T & -R \end{bmatrix} < 0, \quad i = 1, \ldots, L.$$

• NORM-BOUND LDIS: For NLDIs, $\mathcal{E} \supseteq \mathcal{R}_{\rm uce}$ for some state-feedback gain K if there exist $Q > 0$, $\mu > 0$, Y and a diagonal $R > 0$ with unit trace (see LMI (6.19)), such that

$$\begin{bmatrix} QA^T + Y^TB_u^T + AQ + B_uY + \mu B_pB_p^T & QC_q^T + Y^TD_{qu}^T & B_w \\ C_qQ + D_{qu}Y & -\mu I & 0 \\ B_w^T & 0 & -R \end{bmatrix} \leq 0.$$

Remark: For LTI systems and NLDIs, it is possible to eliminate the variable Y.

7.3.3 Reachable sets with unit-peak inputs

For a fixed state-feedback gain K, the set of states reachable with inputs with unit-peak is defined as

$$\mathcal{R}_{\rm up} \triangleq \left\{ x(T) \;\middle|\; \begin{array}{c} x,\ w,\ u \text{ satisfy (7.17)}, \quad u = Kx, \quad x(0)=0, \\ w(t)^T w(t) \le 1, \qquad T \ge 0 \end{array} \right\}.$$

• LTI SYSTEMS: From condition (6.25), $\mathcal{E}$ contains $\mathcal{R}_{\rm uce}$ for the LTI system (7.4) if there exists $\alpha > 0$ such that

$$AQ + QA^T + B_uY + Y^TB_u^T + B_wB_w^T/\alpha + \alpha Q \le 0,$$

where $Y = KQ$.

• POLYTOPIC LDIS: For PLDIs, $\mathcal{E} \supseteq \mathcal{R}_{\rm uce}$ if there exist $Q > 0$ and Y (see condition (6.28)) such that

$$A_iQ + QA_i^T + B_{u,i}Y + Y^TB_{u,i}^T + \alpha Q + B_{w,i}B_{w,i}^T/\alpha \le 0$$

for $i = 1, \ldots, L$.

• NORM-BOUND LDIS: From Chapter 6, $\mathcal{E}$ contains $\mathcal{R}_{\rm uce}$ if there exist $Q > 0$, Y, $\alpha > 0$ and $\mu > 0$ such that

$$\left[\begin{array}{cc} \left(\begin{array}{c} AQ + QA^T + \alpha Q \\ +B_uY + Y^TB_u + \mu B_pB_p^T + B_wB_w^T/\alpha \end{array} \right) & (C_qQ + D_{qu}Y)^T \\ C_qQ + D_{qu}Y & -\mu I \end{array} \right] \le 0.$$

This condition is an LMI for fixed α.

Remark: Again, for LTI systems and NLDIs, it is possible to eliminate the variable Y.

7.4 State-to-Output Properties

We consider state-feedback design to achieve certain desirable state-to-output properties for the LDI (4.5). We set the exogenous input w to zero in (4.5), and consider

$$\dot{x} = A(t)x + B_u(t)u, \qquad z = C_z(t)x.$$

• LTI SYSTEMS: For LTI systems, the state equations are

$$\dot{x} = Ax + B_uu, \qquad z = C_zx + D_{zu}u. \tag{7.21}$$

• POLYTOPIC LDIS: For PLDIs we have

$$\left[\begin{array}{cc} A(t) & B_u(t) \\ C_z(t) & D_{zu}(t) \end{array} \right] \in \mathbf{Co} \left\{ \left[\begin{array}{cc} A_1 & B_{u,1} \\ C_{z,1} & D_{zu,1} \end{array} \right], \ldots, \left[\begin{array}{cc} A_L & B_{u,L} \\ C_{z,L} & D_{zu,L} \end{array} \right] \right\}.$$

• NORM-BOUND LDIs: NLDIs are given by

$$\dot{x} = Ax + B_u u + B_p p, \qquad z = C_z x + D_{zu} u$$
$$q = C_q x + D_{qu} u \qquad p = \Delta(t) q, \qquad \|\Delta(t)\| \le 1,$$

which can be rewritten as

$$\dot{x} = Ax + B_u u + B_p p, \qquad z = C_z x + D_{zu} u$$
$$q = C_q x + D_{qu} u, \qquad p^T p \le q^T q.$$

• DIAGONAL NORM-BOUND LDIs: Finally, DNLDIs are given by

$$\dot{x} = Ax + B_u u + B_p p, \qquad z = C_z x + D_{zu} u, \qquad q = C_q x + D_{qu} u,$$
$$p_i = \delta_i(t) q_i, \qquad |\delta_i(t)| \le 1, \quad i = 1, \ldots, L.$$

which can be rewritten as

$$\dot{x} = Ax + B_u u + B_p p, \qquad z = C_z x + D_{zu} u, \qquad q = C_q x + D_{qu} u,$$
$$|p_i| \le |q_i|, \quad i = 1, \ldots, n_q.$$

7.4.1 Bounds on output energy

We first show how to find a state-feedback gain K such that the output energy (as defined in §6.2.1) of the closed-loop system is less than some specified value. We assume first that the initial condition x_0 is given.

• LTI SYSTEMS: We conclude from LMI (6.39) that for a given state-feedback gain K, the output energy of system (7.21) does not exceed $x_0^T Q^{-1} x_0$, where $Q > 0$ and $Y = KQ$ satisfy

$$\begin{bmatrix} AQ + QA^T + B_u Y + Y^T B_u^T & (C_z Q + D_{zu} Y)^T \\ C_z Q + D_{zu} Y & -I \end{bmatrix} \le 0, \tag{7.22}$$

Regarding Y as a variable, we can then find a state-feedback gain that guarantees an output energy less than γ by solving the LMIP $x_0^T Q^{-1} x_0 \le \gamma$ and (7.22).

Of course, inequality (7.22) is closely related to the classical Linear-Quadratic Regulator (LQR) problem; see the Notes and References.

• POLYTOPIC LDIs: In this case, the output energy is bounded above by $x_0^T Q^{-1} x_0$, where Q satisfies the LMI

$$\begin{bmatrix} \begin{pmatrix} A_i Q + Q A_i^T \\ + B_{u,i} Y + Y^T B_{u,i}^T \end{pmatrix} & (C_{z,i} Q + D_{zu,i} Y)^T \\ C_{z,i} Q + D_{zu,i} Y & -I \end{bmatrix} \le 0, \quad i = 1, \ldots, L$$

for some Y.

• NORM-BOUND LDIs: In the case of NLDIs, the output energy is bounded above by $x_0^T Q^{-1} x_0$, for any $Q > 0$, Y and $\mu \ge 0$ such that

$$\begin{bmatrix} \begin{pmatrix} AQ + QA^T + B_u Y \\ + Y^T B_u^T + \mu B_p B_p^T \end{pmatrix} & (C_z Q + D_{zu} Y)^T & (C_q Q + D_{qu} Y)^T \\ C_z Q + D_{zu} Y & -I & 0 \\ C_q Q + D_{qu} Y & 0 & -\mu I \end{bmatrix} \le 0. \tag{7.23}$$

Remark: As before, we can eliminate the variable Y from this LMI.

• DIAGONAL NORM-BOUND LDIS: Finally, in the case of DNLDIs, the output energy is bounded above by $x_0^T Q^{-1} x_0$, for any $Q > 0$, Y and $M \geq 0$ and diagonal such that

$$\begin{bmatrix} \begin{pmatrix} AQ + QA^T + B_u Y \\ +Y^T B_u^T + B_p M B_p^T \end{pmatrix} & (C_z Q + D_{zu} Y)^T & (C_q Q + D_{qu} Y)^T \\ C_z Q + D_{zu} Y & -I & 0 \\ C_q Q + D_{qu} Y & 0 & -M \end{bmatrix} \leq 0.$$

Given an initial condition, finding a state-feedback gain K so as to minimize the upper bound on the extractable energy for the various LDIs is therefore an EVP. We can extend these results to the case when x_0 is specified to lie in a polytope or an ellipsoid (see §6.2.1). If x_0 is a random variable with $\mathbf{E}\, x_0 x_0^T = X_0$, EVPs that yield state-feedback gains that minimize the expected value of the output energy can be derived.

7.5 Input-to-Output Properties

We finally consider the problem of finding a state-feedback gain K so as to achieve desired properties between the exogenous input w and the output z for the system (7.1). As mentioned in Chapter 6, the list of problems considered here is far from exhaustive.

7.5.1 $\mathbf{L_2}$ and RMS gains

We seek a state-feedback gain K such that the $\mathbf{L_2}$ gain

$$\sup_{\|w\|_2=1} \|z\|_2 = \sup_{\|w\|_2 \neq 0} \frac{\|z\|_2}{\|w\|_2}$$

of the closed-loop system is less than a specified number γ.

• LTI SYSTEMS: As seen in 6.3.2, the $\mathbf{L_2}$ gain for LTI systems is equal to the $\mathbf{H}_\infty$ norm of the corresponding transfer matrix. From §6.3.2, there exists a state-feedback gain K such that the $\mathbf{L_2}$ gain of an LTI system is less than γ, if there exist K and $Q > 0$ such that,

$$\begin{bmatrix} \begin{pmatrix} (A + B_u K)Q + Q(A + B_u K)^T \\ +B_w B_w^T \end{pmatrix} & Q(C_z + D_{zu} K)^T \\ (C_z + D_{zu} K)Q & -\gamma^2 I \end{bmatrix} \leq 0. \qquad (7.24)$$

Introducing $Y = KQ$, this can be rewritten as

$$\begin{bmatrix} AQ + QA^T + B_u Y + Y^T B_u^T + B_w B_w^T & (C_z Q + D_{zu} Y)^T \\ C_z Q + D_{zu} Y & -\gamma^2 I \end{bmatrix} \leq 0. \qquad (7.25)$$

In the case of LTI systems, this condition is necessary and sufficient. Assuming $C_z^T D_{zu} = 0$ and $D_{zu}^T D_{zu}$ invertible, it is possible to simplify the inequality (7.25) and get the equivalent Riccati inequality

$$AQ + QA^T - B_u (D_{zu}^T D_{zu})^{-1} B_u^T + B_w B_w^T + Q C_z^T C_z Q / \gamma^2 \leq 0. \qquad (7.26)$$

The corresponding Riccati equation is readily solved via Hamiltonian matrices. The resulting controller, with state-feedback gain $K = -(D_{zu}^T D_{zu})^{-1} B_u^T$, yields a closed-loop system with $\mathbf{H}_\infty$-norm less than γ.

• POLYTOPIC LDIS: From §6.3.2, there exists a state-feedback gain such that the $\mathbf{L}_2$ gain of a PLDI is less than γ if there exist Y and $Q > 0$ such that

$$\begin{bmatrix} \begin{pmatrix} A_i Q + Q A_i^T + B_{u,i} Y \\ + Y^T B_{u,i}^T + B_{w,i} B_{w,i}^T \end{pmatrix} & (C_{z,i} Q + D_{zu,i} Y)^T \\ C_{z,i} Q + D_{zu,i} Y & -\gamma^2 I \end{bmatrix} \leq 0. \tag{7.27}$$

• NORM-BOUND LDIS: For NLDIs, the LMI that guarantees that the $\mathbf{L}_2$ gain is less than γ for some state-feedback gain K is

$$\begin{bmatrix} \begin{pmatrix} AQ + QA^T \\ + B_u Y + Y^T B_u^T \\ + B_w B_w^T + \mu B_p B_p^T \end{pmatrix} & (C_z Q + D_{zu} Y)^T & (C_q Q + D_{qu} Y)^T \\ C_z Q + D_{zu} Y & -\gamma^2 I & 0 \\ C_q Q + D_{qu} Y & 0 & -\mu I \end{bmatrix} \leq 0. \tag{7.28}$$

We can therefore find state-feedback gains that minimize the upper bound on the $\mathbf{L}_2$ gain, provable with quadratic Lyapunov functions, for the various LDIs by solving EVPs.

7.5.2 Dissipativity

We next seek a state-feedback gain K such that the closed-loop system (7.2) is passive; more generally, assuming $D_{zw}(t)$ is nonzero and square, we wish to maximize the dissipativity, i.e., η satisfying

$$\int_0^T \left(w^T z - \eta w^T w\right) \, dt \geq 0$$

for all $T \geq 0$.

• LTI SYSTEMS: Substituting $Y = KQ$, and using LMI (6.60), we conclude that the dissipation of the LTI system is at least η if the LMI in η, $Q > 0$ and Y holds:

$$\begin{bmatrix} AQ + QA^T + B_u Y + Y^T B_u^T & B_w - QC_z^T - Y^T D_{zu}^T \\ B_w^T - C_z Q - D_{zu} Y & 2\eta I - (D_{zw} + D_{zw}^T) \end{bmatrix} \leq 0.$$

• POLYTOPIC LDIS: From (6.62), there exists a state-feedback gain such that the dissipation exceeds η if there exist $Q > 0$ and Y such that

$$\begin{bmatrix} \begin{pmatrix} A_i Q + Q A_i^T \\ + B_{u,i} Y + Y^T B_{u,i}^T \end{pmatrix} & B_{w,i} - Q C_{z,i}^T - Y^T D_{zu,i} \\ B_{w,i}^T - C_{z,i} Q - D_{zu,i} Y & 2\eta I - (D_{zw,i} + D_{zw,i}^T) \end{bmatrix} \leq 0.$$

• Norm-bound LDIs: From (6.64), there exists a state-feedback gain such that the dissipation exceeds η if there exist $Q > 0$ and Y such that

$$\begin{bmatrix} \begin{pmatrix} AQ + QA^T + B_uY \\ +Y^TB_u^T + \mu B_qB_q^T \end{pmatrix} & \begin{pmatrix} B_w - QC_z^T \\ -Y^TD_{zu}^T \end{pmatrix} & (C_qQ + D_{qu}Y)^T \\ \begin{pmatrix} B_w^T - C_zQ \\ -D_{zu}Y \end{pmatrix} & -(D_{zw} + D_{zw}^T - 2\eta I) & 0 \\ C_qQ + D_{qu}Y & 0 & -\mu I \end{bmatrix} \leq 0$$

We can therefore find state-feedback gains that maximize the lower bound on the dissipativity, provable with quadratic Lyapunov functions, for the various LDIs by solving EVPs.

Remark: As with §6.3.4, we can incorporate scaling techniques into many of the results above to derive componentwise results. Since the new LMIs thus obtained are straightforward to derive, we will omit them here.

7.5.3 Dynamic versus static state-feedback controllers

A *dynamic* state-feedback controller has the form

$$\dot{\bar{x}} = \bar{A}\bar{x} + \bar{B}_y x, \qquad u = \bar{C}_u\bar{x} + \bar{D}_{uy}x, \tag{7.29}$$

where $\bar{A} \in \mathbf{R}^{r\times r}$, $\bar{B}_y \in \mathbf{R}^{r\times n}$, $\bar{C}_u \in \mathbf{R}^{n_u\times r}$, $\bar{D}_{uy} \in \mathbf{R}^{n_u\times n}$. The number r is called the *order* of the controller. Note that by taking $r = 0$, this dynamic state-feedback controller reduces to the state-feedback we have considered so far (which is called *static* state-feedback in this context).

It might seem that the dynamic state-feedback controller (7.29) allows us to meet more specifications than can be met using a static state-feedback. For specifications based on quadratic Lyapunov functions, however, this is not the case. For example, suppose there does not exist a static state-feedback gain that yields closed-loop quadratic stability. In this case we might turn to the more general dynamic state-feedback. We could work out the more complicated LMIs that characterize quadratic stabilizability with dynamic state-feedback. We would find, however, that these more general LMIs are also infeasible. See the Notes and References.

7.6 Observer-Based Controllers for Nonlinear Systems

We consider the nonlinear system

$$\dot{x} = f(x) + B_u u, \qquad y = C_y x, \tag{7.30}$$

where $x : \mathbf{R}_+ \to \mathbf{R}^n$ is the state variable, $u : \mathbf{R}_+ \to \mathbf{R}^p$ is the control variable, and $y : \mathbf{R}_+ \to \mathbf{R}^q$ is the measured or sensed variable. We assume the function $f : \mathbf{R}^n \to \mathbf{R}^n$ satisfies $f(0) = 0$ and

$$\frac{\partial f}{\partial x} \in \mathbf{Co}\,\{A_1, \ldots, A_M\}$$

where $A_1, \ldots, A_L$ are given, which is the same as (4.15).

We look for a stabilizing *observer-based controller* of the form

$$\dot{\bar{x}} = f(\bar{x}) + B_u u + L(C_y\bar{x} - y), \qquad u = K\bar{x}, \tag{7.31}$$

i.e., we search for the matrices K (the *estimated-state feedback gain*) and L (the *observer gain*) such that the closed-loop system

$$\begin{bmatrix} \dot{x} \\ \dot{\bar{x}} \end{bmatrix} = \begin{bmatrix} f(x) + B_u K\bar{x} \\ -LC_y x + f(\bar{x}) + (B_u K + LC_y)\bar{x} \end{bmatrix} \tag{7.32}$$

is stable. The closed-loop system (7.32) is stable if it is quadratically stable, which is true if there exists a positive-definite matrix $\tilde{P} \in \mathbf{R}^{2n\times 2n}$ such that for any nonzero trajectory x, $\bar{x}$, we have:

$$\frac{d}{dt}\begin{bmatrix} x \\ \bar{x} \end{bmatrix}^T \tilde{P} \begin{bmatrix} x \\ \bar{x} \end{bmatrix} < 0.$$

In the Notes and References, we prove that this is true if there exist P, Q, Y, and W such that the LMIs

$$Q > 0, \quad A_i Q + QA_i^T + B_u Y + Y^T B_u^T < 0, \quad i = 1,\ldots,M \tag{7.33}$$

and

$$P > 0, \quad A_i^T P + PA_i + WC_y + C_y^T W^T < 0, \quad i = 1,\ldots,M \tag{7.34}$$

hold. To every P, Q, Y, and W satisfying these LMIs, there corresponds a stabilizing observer-based controller of the form (7.31), obtained by setting $K = YQ^{-1}$ and $L = P^{-1}W$.

Using the elimination procedure of §2.6.2, we can obtain equivalent conditions in which the variables Y and W do not appear. These conditions are that some $P > 0$ and $Q > 0$ satisfy

$$A_i Q + QA_i^T < \sigma B_u B_u^T, \quad i = 1,\ldots,M \tag{7.35}$$

and

$$A_i^T P + PA_i < \mu C_y^T C_y, \quad i = 1,\ldots,M \tag{7.36}$$

for some σ and μ. By homogeneity we can freely set $\sigma = \mu = 1$. For any $P > 0$, $Q > 0$ satisfying these LMIs with $\sigma = \mu = 1$, we can obtain a stabilizing observer-based controller of the form (7.31), by setting $K = -(1/2)B_u^T Q^{-1}$ and $L = -(1/2)P^{-1}C_y$.

Another equivalent condition is that

$$\tilde{B}_u^T \left(A_i Q + QA_i^T\right) \tilde{B}_u < 0, \quad i = 1,\ldots,M$$

and

$$\tilde{C}_y \left(A_i^T P + PA_i\right) \tilde{C}_y^T < 0, \quad i = 1,\ldots,M$$

hold for some $P > 0$, $Q > 0$, where $\tilde{B}_u$ and $\tilde{C}_y^T$ are orthogonal complements of B_u and C_y^T, respectively. If P and Q satisfy these LMIs, then they satisfy the LMIs (7.36) and (7.35) for some σ and μ. The observer-based controller with $K = -(\sigma/2)B_u^T Q^{-1}$ and $L = -(\mu/2)P^{-1}C_y$ stabilizes the nonlinear system (7.30).

Notes and References

Lyapunov functions and state-feedback

The extension of Lyapunov's methods to the state-feedback synthesis problem has a long history; see e.g., [BAR70A, LEF65]. It is clearly related to the theory of optimal control, in which the natural Lyapunov function is the Bellman–Pontryagin "min-cost-to-go" or value function (see [PON61, PON62]). We suspect that the methods described in this chapter can be interpreted as a search for suboptimal controls in which the candidate value functions are restricted to a specific class, i.e., quadratic. In the general optimal control problem, the (exact) value function satisfies a partial differential equation and is hard to compute; by restricting our search to quadratic approximate value functions we end up with a low complexity task, i.e., a convex problem involving LMIs.

Several approaches use Lyapunov-like functions to synthesize nonlinear state-feedback control laws. One example is the "Lyapunov Min-Max Controller" described in [GUT79, GP82, COR85, CL90, WEI94]. See also [ZIN90, SC91, RYA88, GR88].

Quadratic stabilizability

The term "quadratic stabilizability" seems to have been coined by Hollot and Barmish in [HB80], where the authors give necessary and sufficient conditions for it; see also [LEI79, GUT79, BCL83, PB84, COR85]. Petersen [PET85], shows that there exist LDIs that are quadratically stabilizable, though not via linear state-feedback; see also [SP94E]. Hollot [HOL87] describes conditions under which a quadratically stabilizable LDI has infinite stabilizability margin; see also [SRC93]. Wei [WEI90] derives necessary and sufficient "sign-pattern" conditions on the perturbations for the existence of a linear feedback for quadratic stability for a class of single-input uncertain linear dynamical systems. Petersen [PET87B] derives a Riccati-based approach for stabilizing NLDIs, which is in fact the state-feedback $\mathbf{H}_\infty$ equation (7.28) (this is shown in [PET87A]).

The change of variables $Q = P^{-1}$ and $Y = KP^{-1}$, which enables the extension of many of the results of Chapters 5 and 6 to state-feedback synthesis, is due to Bernussou, Peres and Geromel [BPG89A]. See also [GPB91, BPG89B, BPG89A, PBG89, PGB93]. This change of variables, though not explicitly described, is used in the paper by Thorp and Barmish [TB81]; see also [HB80, BAR83, BAR85]. For a recent and broad review about quadratic stabilization of uncertain systems, see [COR94].

In [SON83], Sontag formulates a general theorem giving necessary and sufficient Lyapunov type conditions for stabilizability. For LTI systems his criteria reduce to the LMIs in this chapter.

Quadratic Lyapunov functions for proving performance

The addition of performance considerations in the analysis of LDIs can be traced back to 1955 and even earlier, when Letov [LET61] studied the performance of unknown, nonlinear and possibly time-varying control systems and calls it "the problem of control quality". At that time, the performance criteria were decay rate and output peak deviations for systems subject to bounded peak inputs (see [LET61, P.234] for details).

In the case of LTI systems, adding performance indices allows one to recover many classical results of automatic control. The problem of maximizing decay rate is discussed in great detail in [YAN92]; other references on decay rates and quadratic stability margins are [GU92B, GU92A, GPB91, GU92B, SKO91B, PZP+92, EBFB92, COR90]. Arzelier et al. [ABG93] consider the problem of robust stabilization with pole assignment for PLDIs via cutting-plane techniques.

The problem of approximating reachable sets for LTI systems under various assumptions on the energy of the exogenous input has been considered for example by Skelton [SZ92] and references therein; see also the Notes and References of the previous chapter. See also [OC87].

The idea that the output variance for white noise inputs (and related LQR-like performance measures) can be minimized using LMIs can be found in [BH88A, BH89, RK91, KR91, KKR93, BB91, FBBE92, RK93, SI93]. Peres, Souza and Geromel [PSG92, PG93] consider the extension to PLDIs. The counterpart for NLDIs is in Petersen, McFarlane and Rotea [PM92, PMR93], as well as in Stoorvogel [STO91].

The problem of finding a state-feedback gain to minimize the scaled $\mathbf{L}_2$ gain for LTI systems is discussed in [EBFB92]. The $\mathbf{L}_2$ gain of PLDIs has been studied by Obradovic and

Valavani [OV92], by Peres, Geromel and Souza [PGS91] and also by Ohara, Masubuchi and Suda [OMS93]. NLDIs have been studied in this context by Petersen [PET89], and DeSouza et al. [XFDS92, XDS92, WXDS92, DSFX93], who point out explicitly that quadratic stability of NLDIs with an $\mathbf{L}_2$ gain bound is equivalent to the scaled $\mathbf{H}_\infty$ condition (6.54); see also [ZKSN92, GU93]. For LTI systems, it is interesting to compare the inequality (7.25) which provides a necessary and sufficient condition for a system to have $\mathbf{L}_2$ gain less than γ with the corresponding matrix inequality found in the article by Stoorvogel and Trentelman [ST90]. In particular, the quadratic matrix inequality found there (expressed in our notation) is

$$\begin{bmatrix} A^TP + PA + \gamma^2 PB_wB_w^TP + C_z^TC_z & PB_u + C_z^TD_{zu} \\ B_u^TP + D_{zu}^TC_z & D_{zu}^TD_{zu} \end{bmatrix} \geq 0,$$

which does not possess any obvious convexity property. We note that the Riccati inequality (7.26) is also encountered in full-state feedback $\mathbf{H}_\infty$ control [PET87A]. Its occurrence in control theory is much older;; it appears verbatim in some articles on the theory of nonzero-sum differential games; see for example Starr and Ho [SH68], Yakubovich [YAK70, YAK71] and Mageirou [MAG76].

Kokotovic [KOK92] considers stabilization of nonlinear systems, and provides a motivation for rendering an LTI system passive via state-feedback (see §6.3.3). In [PP92], the authors give analytic conditions for making an LTI system passive via state-feedback; see also [SKS93].

Finally, we mention an instance of a Lyapunov function with a built-in performance criterion. In the book by Aubin and Cellina [AC84, §6], we find:

> We shall investigate whether differential inclusions $\dot{x} \in F(x(t))$, $x(0) = x_0$ have trajectories satisfying the property
>
> $$\forall t > s, \quad V(x(t)) - V(x(s)) + \int_s^t W(x(\tau), \dot{x}(\tau))\, d\tau \leq 0. \tag{7.37}$$
>
> We shall then say that a function $V[\cdots]$ satisfying this condition is a Lyapunov function for F with respect to W.

Similar ideas can be found in [CP72]. We shall encounter such Lyapunov functions in §8.2. Inequality (7.37) is often called a "dissipation inequality" [WIL71B].

The problem of minimizing the control effort given a performance bound on the closed-loop system can be found in the articles of Schömig, Sznaier and Ly [SSL93], and Grigoriadis, Carpenter, Zhu and Skelton [GCZS93].

LMI formulation of LQR problem

For the LTI system $\dot{x} = Ax + B_uu$, $z = C_zx + D_{zu}u$, the Linear-Quadratic Regulator (LQR) problem is: Given an initial condition $x(0)$, find the control input u that minimizes the output energy $\int_0^\infty z^Tz\,dt$. We assume for simplicity that (A, B, C) is minimal, $D_{zu}^TD_{zu}$ is invertible and $D_{zu}^TC_z = 0$.

It turns out that the optimal input u can be expressed as a constant state-feedback $u = Kx$, where $K = -(D_{zu}^TD_{zu})^{-1}B_u^TP_{\rm are}$ and $P_{\rm are}$ is the unique positive-definite matrix that satisfies the algebraic Riccati equation

$$A^TP_{\rm are} + P_{\rm are}A - P_{\rm are}B_u(D_{zu}^TD_{zu})^{-1}B_u^TP_{\rm are} + C_z^TC_z = 0. \tag{7.38}$$

The optimal output energy for an initial condition $x(0)$ is then given by $x(0)^TP_{\rm are}x(0)$. With $Q_{\rm are} = P_{\rm are}^{-1}$, we can rewrite (7.38) as

$$AQ_{\rm are} + Q_{\rm are}A^T - B_u(D_{zu}^TD_{zu})^{-1}B_u^T + Q_{\rm are}C_z^TC_zQ_{\rm are} = 0, \tag{7.39}$$

and the optimal output energy as $x(0)^T Q_{\rm are}^{-1} x(0)$.

In section §7.4.1, we considered the closely related problem of finding a state-feedback gain K that minimizes an upper bound on the energy of the output, given an initial condition. For LTI systems, the upper bound equals the output energy, and we showed that in this case, the minimum output energy was given by minimizing $x(0)^T Q^{-1} x(0)$ subject to $Q > 0$ and

$$\begin{bmatrix} AQ + QA^T + B_uY + Y^TB_u^T & (C_zQ + D_{zu}Y)^T \\ C_zQ + D_{zu}Y & -I \end{bmatrix} \le 0. \tag{7.40}$$

Of course, the optimal value of this EVP must equal the optimal output energy given via the solution to the Riccati solution (7.39); the Riccati equation can be thus be interpreted as yielding an analytic solution to the EVP. We can derive this analytic solution for the EVP via the following steps. First, it can be shown, using a simple completion-of-squares argument that LMI (7.40) holds for some $Q > 0$ and Y if and only if it holds for $Y = -(D_{zu}^T D_{zu})^{-1} B_u^T$; in this case, $Q > 0$ must satisfy

$$QA^T + AQ + QC_z^TC_zQ - B_u\left(D_{zu}^TD_{zu}\right)^{-1}B_u^T \le 0. \tag{7.41}$$

Next, it can be shown by standard manipulations that if $Q > 0$ satisfies (7.41), then $Q \le Q_{\rm are}$. Therefore for every initial condition $x(0)$, we have $x(0)^T Q_{\rm are}^{-1} x(0) \le x(0)^T Q^{-1} x(0)$, and therefore the optimal value of the EVP is just $x(0)^T Q_{\rm are}^{-1} x(0)$, and the optimal state-feedback gain is $K_{\rm opt} = Y_{\rm opt} Q_{\rm are}^{-1} = -(D_{zu}^T D_{zu})^{-1} B_u^T P_{\rm are}$.

It is interesting to note that with $P = Q^{-1}$, the EVP is equivalent to minimizing $x(0)^T P x(0)$ subject to

$$A^TP + PA + C_z^TC_z - PB_u\left(D_{zu}^TD_{zu}\right)^{-1}B_u^TP \le 0, \tag{7.42}$$

which is *not* a convex constraint in P. However, the problem of *maximizing* $x(0)^T P x(0)$ subject $P > 0$ and the constraint

$$A^TP + PA - PB_u(D_{zu}^TD_{zu})^{-1}B_u^TP + C_zC_z^T \ge 0,$$

which is nothing other than (7.42), but with the inequality reversed, is an EVP; it is well-known (see for example [WIL71B]) that this EVP is another formulation of the LQR problem.

Static versus dynamic state-feedback

An LTI system can be stabilized using dynamic state-feedback if and only if it can be stabilized using static state-feedback (see for example, [KAI80]). In fact, similar statements can be made for all the properties and for all the LDIs considered in this chapter, provided the properties are specified using quadratic Lyapunov functions. We will demonstrate this fact on one simple problem considered in this chapter.

Consider the LTI system

$$\dot{x} = Ax + B_uu + B_ww, \quad x(0) = 0, \qquad z = C_zx. \tag{7.43}$$

From §7.5.1, the exists a static state-feedback $u = Kx$ such that the $\mathbf{L}_2$ gain from w to z does not exceed γ if the LMI in variables $Q > 0$ and Y is feasible:

$$\begin{bmatrix} AQ + QA^T + B_uY + Y^TB_u^T + B_wB_w^T & QC_z^T \\ C_zQ & -\gamma^2 I \end{bmatrix} \le 0.$$

(In fact this condition is also necessary, but this is irrelevant to the current discussion.) Eliminating Y, we get the equivalent LMI in $Q > 0$ and scalar σ:

$$AQ + QA^T + B_w B_w^T + \frac{QC_z^T C_z Q}{\gamma^2} \leq \sigma B_u B_u^T. \tag{7.44}$$

We next consider a dynamic state-feedback for the system:

$$\dot{x}_c = A_c x_c + B_c x, \quad x_c(0) = 0, \qquad u = C_c x_c + D_c x. \tag{7.45}$$

We will show that there exist matrices A_c, B_c, C_c and D_c such that the $\mathbf{L}_2$ gain of the system (7.43,7.45) does not exceed γ if and only if the LMI (7.44) holds. This will establish that the smallest upper bound on the $\mathbf{L}_2$ gain, provable via quadratic Lyapunov functions, is the same irrespective of whether a static state-feedback or a dynamic state-feedback is employed.

We will show only the part that is not obvious. With state vector $x_{\rm cl} = [x^T \ \ x_c^T]^T$, the closed-loop system is

$$\dot{x}_{\rm cl} = A_{\rm cl} x_{\rm cl} + B_{\rm cl} u, \quad x_{\rm cl}(0) = 0, \qquad z = C_{\rm cl} x_{\rm cl},$$

where $A_{\rm cl} = A_{\rm big} + B_{\rm big} K_{\rm big}$ with

$$A_{\rm big} = \begin{bmatrix} A & 0 \\ 0 & 0 \end{bmatrix}, \qquad B_{\rm big} = \begin{bmatrix} 0 & B_u \\ I & 0 \end{bmatrix}, \qquad K_{\rm big} = \begin{bmatrix} B_c & A_c \\ D_c & C_c \end{bmatrix},$$

$$B_{\rm cl} = \begin{bmatrix} B_w \\ 0 \end{bmatrix}, \qquad C_{\rm cl} = [C_z \ \ 0].$$

From §6.3.2, the $\mathbf{L}_2$ gain of the closed-loop system does not exceed γ if the LMI in $P_{\rm big} > 0$ is feasible:

$$\begin{bmatrix} A_{\rm big}^T P + P_{\rm big} A + PB_{\rm big} K_{\rm big} + K_{\rm big}^T B_{\rm big}^T P + C_{\rm cl}^T C_{\rm cl} & PB_{\rm cl} \\ B_{\rm cl}^T P & -\gamma^2 I \end{bmatrix} \leq 0.$$

Eliminating $K_{\rm big}$ from this LMI yields two equivalent LMIs in $P_{\rm big} > 0$ and a scalar τ.

$$\begin{bmatrix} A_{\rm big}^T P + P_{\rm big} A - \tau I + C_{\rm cl}^T C_{\rm cl} & PB_{\rm cl} \\ B_{\rm cl}^T P & -\gamma^2 I \end{bmatrix} \leq 0,$$

$$\begin{bmatrix} A_{\rm big}^T P + P_{\rm big} A - \tau PB_{\rm big} B_{\rm big}^T P + C_{\rm cl}^T C_{\rm cl} & PB_{\rm cl} \\ B_{\rm cl}^T P & -\gamma^2 I \end{bmatrix} \leq 0.$$

It is easy to verify that the first LMI holds for large enough τ, while the second reduces to $\tau > 0$ and

$$AQ_{11} + Q_{11} A^T + \frac{B_w B_w^T}{\gamma^2} + Q_{11} C_z C_z^T Q_{11} \leq \tau B_u B_u^T,$$

where Q_{11} is the leading $n \times n$ block of $P_{\rm big}^{-1}$. The last LMI is precisely (7.44) with the change of variables $Q = \gamma^2 Q_{11}$ and $\sigma = \gamma^2 \tau$.

A similar result is established [PZP+92]; other references that discuss the question of when dynamic or nonlinear feedback does better than static feedback include Petersen, Khargonekar and Rotea [PET85, KPR88, RK88, KR88, PET88, RK89].

Dynamic output-feedback for LDIs

A natural extension to the results of this chapter is the synthesis of output-feedback to meet various performance specifications. We suspect, however, that output-feedback synthesis problems have high complexity and are therefore unlikely to be recast as LMI problems. Several researchers have considered such problems: Steinberg and Corless [SC85], Galimidi and Barmish [GB86], and Geromel, Peres, de Souza and Skelton [PGS93]; see also [HCM93].

One variation of this problem is the "reduced-order controller problem", that is, finding dynamic output-feedback controllers with the smallest possible number of states. It is shown in [PZPB91, EG93, PAC94] that this problem can be reduced to the problem of minimizing the rank of a matrix $P \geq 0$ subject to certain LMIs. The general problem of minimizing the rank of a matrix subject to an LMI constraint has high complexity; in fact, a special rank minimization problem can be shown to be equivalent to the NP-hard zero-one linear programming problem (see e.g. [DAV94]). Nevertheless several researchers have tried heuristic, local algorithms for such problems and report practical success. Others studying this topic include Iwasaki, Skelton, Geromel and de Souza [IS93B, ISG93, GDSS93, SIG93].

Observer-based controllers for nonlinear systems

We prove that the LMIs (7.33) and (7.34) (in the variables $P > 0$, $Q > 0$, W and Y) ensure the existence of an observer-based controller of the form (7.31) which makes the system (7.30) quadratically stable.

We start by rewriting the system (7.32) in the coordinates $\bar{x}$, $\bar{x} - x$:

$$\begin{bmatrix} \dot{\bar{x}} \\ \dot{\bar{x}} - \dot{x} \end{bmatrix} = \begin{bmatrix} LC_y(\bar{x} - x) + f(\bar{x}) + B_u K\bar{x} \\ f(\bar{x}) - f(x) + LC_y(\bar{x} - x) \end{bmatrix}. \tag{7.46}$$

Using the results of §4.3, we can write

$$\begin{bmatrix} \dot{\bar{x}} \\ \dot{\bar{x}} - \dot{x} \end{bmatrix} \in \mathbf{Co}\left\{ \tilde{A}_{ij} \,\middle|\, 1 \leq i \leq M, \quad 1 \leq j \leq M \right\} \begin{bmatrix} \bar{x} \\ \bar{x} - x \end{bmatrix}$$

with

$$\tilde{A}_{ij} = \begin{bmatrix} A_i + B_u K & LC_y \\ 0 & A_j + LC_y \end{bmatrix}, \quad i, j = 1, \ldots, M.$$

Simple state-space manipulations show the system (7.32) is quadratically stable if and only if the system (7.46) is quadratically stable. Therefore, the system (7.32) is quadratically stable if there exists $\tilde{P} > 0$ such that

$$\tilde{A}_{ij}^T \tilde{P} + \tilde{P}\tilde{A}_{ij} < 0, \quad i, j = 1, \ldots, M. \tag{7.47}$$

We will now show that the LMI (7.47) has a solution if and only if the LMIs (7.33) and (7.34) do.

First suppose that (7.47) has a solution $\tilde{P}$. With $\tilde{P}$ partitioned as $n \times n$ blocks,

$$\tilde{P} = \begin{bmatrix} P_{11} & P_{12} \\ P_{12}^T & P_{22} \end{bmatrix},$$

we easily check that the inequality (7.47) implies

$$(A_i + B_u K)^T P_{11} + P_{11}(A_i + B_u K) < 0, \quad i = 1, \ldots, M. \tag{7.48}$$

With the change of variables $Q = P_{11}^{-1}$ and $Y = B_u Q$, the inequality (7.48) is equivalent to (7.33).

Define $\tilde{Q} = \tilde{P}^{-1}$. The inequality (7.47) becomes

$$\tilde{A}_{ij}\tilde{Q} + \tilde{Q}\tilde{A}_{ij}^T < 0, \quad i,j = 1,\ldots,M.$$

Denoting by Q_{22} the lower-right $n \times n$ block of $\tilde{Q}$, we see that this inequality implies

$$(A_i + LC_y)Q_{22} + Q_{22}(A_i + LC_y)^T < 0, \quad i = 1,\ldots,M. \tag{7.49}$$

With $P = Q_{22}^{-1}$ and $W = PL$, the inequality (7.49) is equivalent to (7.34).

Conversely, suppose that P, Q, Y and W satisfy the LMIs (7.34) and (7.33) and define $L = P^{-1}W$, $K = YQ^{-1}$. We now prove there exists a positive λ such that

$$\tilde{P} = \begin{bmatrix} \lambda Q^{-1} & 0 \\ 0 & P \end{bmatrix}$$

satisfies (7.47) and therefore proves the closed-loop system (7.32) is quadratically stable. We compute

$$\tilde{A}_{ij}^T\tilde{P} + \tilde{P}\tilde{A}_{ij} = \begin{bmatrix} \lambda \begin{pmatrix} (A_i + B_uK)^TQ^{-1} \\ +Q^{-1}(A_i + B_uK) \end{pmatrix} & \lambda Q^{-1}LC_y \\ \lambda(LC_y)^TQ^{-1} & \begin{array}{c}(A_j + LC_y)^TP \\ +P(A_j + LC_y)\end{array} \end{bmatrix},$$

for $i,j = 1,\ldots,M$. Therefore, using Schur complements, the closed-loop system (7.32) is quadratically stable if $\lambda > 0$ satisfies

$$\begin{array}{r}\lambda(Q^{-1}LC_y((A_j + LC_y)^TP + P(A_j + LC_y))^{-1}(LC_y)^TQ^{-1}) \\ -(A + B_uK)^TQ^{-1} - Q^{-1}(A + B_uK) \quad > \quad 0\end{array}$$

for $i,j = 1,\ldots,M$. This condition is satisfied for any $\lambda > 0$ such that

$$\lambda \min_{1\le j\le M} \mu_j > \max_{1\le i\le M} \nu_i,$$

where

$$\mu_j = \lambda_{\min}(Q^{-1}LC_y((A_j + LC_y)^TP + P(A_j + LC_y))^{-1}(LC_y)^TQ^{-1}), \quad j = 1,\ldots,M$$

and

$$\nu_i = \lambda_{\max}((A_i + B_uK)^TQ^{-1} + Q^{-1}(A_i + B_uK)), \quad i = 1,\ldots,M.$$

Since (7.34) and (7.33) are satisfied, such a λ always exists.

Problems of observer-based controller design along with quadratic Lyapunov functions that prove stability have been considered by Bernstein and Haddad [BH93] and Yaz [YAZ93].

Gain-scheduled or parameter-dependent controllers

Several researchers have extended the results on state-feedback synthesis to handle the case in which the controller parameters can depend on system parameters. Such controllers are called gain-scheduled or parameter-dependent. We refer the reader to the articles cited for precise explanations of what these control laws are, and which design problems can be recast as LMI problems. Lu, Zhou and Doyle in [LZD91] and Becker and Packard [BEC93, BP91] consider the problem in the context of NLDIs. Other relevant references include [PB92, PAC94, PBPB93, BPPB93, GA94, IS93A, IS93B, AGB94].

Chapter 8

Lur'e and Multiplier Methods

8.1 Analysis of Lur'e Systems

We consider the *Lur'e system*

$$\begin{array}{ll} \dot{x} = Ax + B_p p + B_w w, & q = C_q x, \\ z = C_z x, & p_i(t) = \phi_i(q_i(t)), \quad i = 1, \ldots, n_p, \end{array} \tag{8.1}$$

where $p(t) \in \mathbf{R}^{n_p}$, and the functions ϕ_i satisfy the $[0,1]$ sector conditions

$$0 \le \sigma\phi_i(\sigma) \le \sigma^2 \quad \text{for all} \quad \sigma \in \mathbf{R}, \tag{8.2}$$

or, equivalently,

$$\phi_i(\sigma)(\phi_i(\sigma) - \sigma) \le 0 \quad \text{for all} \quad \sigma \in \mathbf{R}.$$

The data in this problem are the matrices A, B_p, B_w, C_q and C_z. The results that follow will hold for any nonlinearities ϕ_i satisfying the sector conditions. In cases where the ϕ_i are known, however, the results can be sharpened.

It is possible to handle the more general sector conditions

$$\alpha_i\sigma^2 \le \sigma\phi_i(\sigma) \le \beta_i\sigma^2 \quad \text{for all} \quad \sigma \in \mathbf{R},$$

where α_i and β_i are given. Such systems are readily transformed to the form given in (8.1) and (8.2) by a *loop transformation* (see the Notes and References).

An important special case of (8.1), (8.2) occurs when the functions ϕ_i are linear, i.e., $\phi_i(\sigma) = \delta_i\sigma$, where $\delta_i \in [0,1]$. In the terminology of control theory, this is referred to as a system with unknown-but-constant parameters. It is important to distinguish this case from the PLDI obtained with (8.1), $\phi_i(\sigma) = \delta_i(t)\sigma$ and $\delta_i(t) \in [0,1]$, which is referred to as a system with unknown, time-varying parameters.

Our analysis will be based on Lyapunov functions of the form

$$V(\xi) = \xi^T P \xi + 2\sum_{i=1}^{n_p} \lambda_i \int_0^{C_{i,q}\xi} \phi_i(\sigma)\, d\sigma, \tag{8.3}$$

where $C_{i,q}$ denotes the ith row of C_q. Thus the data describing the Lyapunov function are the matrix P and the scalars λ_i, $i = 1, \ldots, n_p$. We require $P > 0$ and $\lambda_i \ge 0$, which implies that $V(\xi) \ge \xi^T P \xi > 0$ for nonzero ξ.

Note that the Lyapunov function (8.3) depends on the (not fully specified) nonlinearities ϕ_i. Thus the data P and λ_i should be thought of as providing a *recipe* for constructing a specific Lyapunov function given the nonlinearities ϕ_i. As an example, consider the special case of a system with unknown-but-constant parameters, i.e., $\phi_i(\sigma) = \delta_i \sigma$. The Lyapunov function will then have the form

$$V(\xi) = \xi^T \left(P + C_q^T \Delta \Lambda C_q\right) \xi$$

where $\Delta = \mathbf{diag}(\delta_1, \ldots, \delta_{n_p})$ and $\Lambda = \mathbf{diag}(\lambda_1, \ldots, \lambda_{n_p})$. In other words, we are really synthesizing a parameter-dependent quadratic Lyapunov function for our parameter-dependent system.

8.1.1 Stability

We set the exogenous input w to zero, and seek P and λ_i such that

$$\frac{dV(x)}{dt} < 0 \text{ for all nonzero } x \text{ satisfying (8.1) and (8.2).} \tag{8.4}$$

Since

$$\frac{dV(x)}{dt} = 2\left(x^T P + \sum_{i=1}^{n_p} \lambda_i p_i C_{i,q}\right)(Ax + B_p p),$$

condition (8.4) holds if and only if

$$\left(\xi^T P + \sum_{i=1}^{n_p} \lambda_i \pi_i C_{i,q}\right)(A\xi + B_p \pi) < 0$$

for all nonzero ξ satisfying

$$\pi_i(\pi_i - c_{i,q}\xi) \le 0, \quad i = 1, \ldots, n_p. \tag{8.5}$$

It is easily shown that

$$\{(\xi,\pi) \mid \xi \ne 0, \quad (8.5)\} = \{(\xi,\pi) \mid \xi \ne 0 \text{ or } \pi \ne 0, \quad (8.5)\}.$$

The $\mathcal{S}$-procedure then yields the following sufficient condition for (8.4): The LMI in $P > 0$, $\Lambda = \mathbf{diag}(\lambda_1, \ldots, \lambda_{n_p}) \ge 0$ and $T = \mathbf{diag}(\tau_1, \ldots, \tau_{n_p}) \ge 0$

$$\begin{bmatrix} A^T P + PA & PB_p + A^T C_q^T \Lambda + C_q^T T \\ B_p^T P + \Lambda C_q A + T C_q & \Lambda C_q B_p + B_p^T C_q^T \Lambda - 2T \end{bmatrix} < 0 \tag{8.6}$$

holds.

Remark: When we set $\Lambda = 0$ we obtain the LMI

$$\begin{bmatrix} A^T P + PA & PB_p + C_q^T T \\ B_p^T P + T C_q & -2T \end{bmatrix} < 0,$$

which can be interpreted as a condition for the existence of a quadratic Lyapunov function for the Lur'e system, as follows. Perform a loop transformation on the Lur'e system to bring the nonlinearities to sector $[-1, 1]$. Then the LMI above is the same (to within a scaling of the variable T) as one that yields quadratic stability of the associated DNLDI; see §5.1.

Remark: Condition (8.6) is only a sufficient condition for the existence of a Lur'e Lyapunov function that proves stability of system (8.1). It is also necessary when there is only one nonlinearity, i.e., when $n_p = 1$.

8.1.2 Reachable sets with unit-energy inputs

We consider the set reachable with unit-energy inputs,

$$\mathcal{R}_{\rm ue} \stackrel{\Delta}{=} \left\{ x(T) \;\middle|\; \begin{array}{ll} x,\, w \text{ satisfy (8.3)}, & x(0) = 0 \\ \displaystyle\int_0^T w^T w\, dt \le 1, & T \ge 0 \end{array} \right\}$$

The set

$$\mathcal{F} = \left\{ \xi \;\middle|\; \xi^T P \xi + 2 \sum_{i=1}^{n_p} \lambda_i \int_0^{C_{i,q}\xi} \phi_i(\sigma)\, d\sigma \le 1 \right\}$$

contains $\mathcal{R}_{\rm ue}$ if

$$\frac{d}{dt} V(x) \le w^T w \quad \text{for all } x \text{ and } w \text{ satisfying (8.1) and (8.2).} \tag{8.7}$$

From the $\mathcal{S}$-procedure, the condition (8.7) holds if there exists $T = \mathbf{diag}(\tau_1, \ldots, \tau_{n_p}) \ge 0$ such that

$$\left[\begin{array}{ccc} A^T P + PA & PB_p + A^T C_q^T \Lambda + C_q^T T & PB_w \\ B_p^T P + \Lambda C_q A + T C_q & \Lambda C_q B_p + B_p^T C_q^T \Lambda - 2T & \Lambda C_q B_w \\ B_w^T P & B_w^T C_q^T \Lambda & -I \end{array} \right] \le 0 \tag{8.8}$$

holds, where $\Lambda = \mathbf{diag}(\lambda_1, \ldots, \lambda_{n_p}) \ge 0$.

Of course, the ellipsoid $\mathcal{E} = \{\xi \in \mathbf{R}^n \mid \xi^T P \xi \le 1\}$ contains $\mathcal{F}$, so that $\mathcal{E} \supseteq \mathcal{F} \supseteq \mathcal{R}_{\rm ue}$. Therefore, $\mathcal{E}$ gives an outer approximation of $\mathcal{R}_{\rm ue}$ if the nonlinearities are not known. If the nonlinearities ϕ_i are known, $\mathcal{F}$ gives a possibly better outer approximation. We can use these results to prove that a point x_0 does not belong to the reachable set by solving appropriate LMIPs.

8.1.3 Output energy bounds

We consider the system (8.1) with initial condition $x(0)$, and compute upper bounds on the output energy

$$J = \int_0^\infty z^T z\, dt$$

using Lyapunov functions V of the form (8.3).

If

$$\frac{d}{dt} V(x) + z^T z \le 0 \qquad \text{for all } x \text{ satisfying (8.1),} \tag{8.9}$$

then $J \le V(x(0))$. The condition (8.9) is equivalent to

$$2 \left(\xi^T P + \sum_{i=1}^{n_p} \lambda_i \pi_i C_{i,q} \right) (A\xi + B_p \pi) + \xi^T C_z^T C_z \xi \le 0,$$

for every ξ satisfying

$$\pi_i (\pi_i - c_{i,q}\xi) \le 0, \quad i = 1, \ldots, n_p.$$

Using the $\mathcal{S}$-procedure, we conclude that the condition (8.9) holds if there exists $T = \mathbf{diag}(\tau_1, \ldots, \tau_{n_p}) \geq 0$ such that

$$\begin{bmatrix} A^TP + PA + C_z^TC_z & PB_p + A^TC_q^T\Lambda + C_q^TT \\ B_p^TP + \Lambda C_qA + TC_q & \Lambda C_qB_p + B_p^TC_q^T\Lambda - 2T \end{bmatrix} \leq 0. \tag{8.10}$$

Since $V(x(0)) \leq x(0)^T(P + C_q^T\Lambda C_q)x(0)$, an upper bound on J is obtained by solving the following EVP in the variables P, $\Lambda = \mathbf{diag}(\lambda_1, \ldots, \lambda_{n_p})$ and $T = \mathbf{diag}(\tau_1, \ldots, \tau_{n_p})$:

$$\begin{array}{ll} \text{minimize} & x(0)^T\left(P + C_q^T\Lambda C_q\right)x(0) \\ \\ \text{subject to} & (8.10),\ T \geq 0,\ \Lambda \geq 0,\ P > 0 \end{array}$$

If the nonlinearities ϕ_i are known, we can obtain possibly better bounds on J by modifying the objective in the EVP appropriately.

8.1.4 $\mathbf{L}_2$ gain

We assume that $D_{zw} = 0$ for simplicity. If there exists a Lyapunov function of the form (8.3) and $\gamma \geq 0$ such that

$$\frac{d}{dt}V(x) \leq \gamma^2w^Tw - z^Tz \qquad \text{for all } x \text{ and } w \text{ satisfying (8.1)}, \tag{8.11}$$

then the $\mathbf{L}_2$ gain of the system (8.1) does not exceed γ. The condition (8.11) is equivalent to

$$2\left(\xi^TP + \sum_{i=1}^{n_p}\lambda_i\pi_iC_{i,q}\right)(A\xi + B_p\pi) \leq \gamma^2w^Tw - \xi^TC_z^TC_z\xi$$

for any ξ satisfying $\pi_i(\pi_i - c_{i,q}\xi) \leq 0, \quad i = 1, \ldots, n_p$. Using the $\mathcal{S}$-procedure, this is satisfied if there exists $T = \mathbf{diag}(\tau_1, \ldots, \tau_{n_p}) \geq 0$ such that

$$\begin{bmatrix} A^TP + PA + C_z^TC_z & PB_p + A^TC_q^T\Lambda + C_q^TT & PB_w \\ B_p^TP + \Lambda C_qA + TC_q & \Lambda C_qB_p + B_p^TC_q^T\Lambda - 2T & \Lambda C_qB_w \\ B_w^TP & B_w^TC_q^T\Lambda & -\gamma^2I \end{bmatrix} \leq 0. \tag{8.12}$$

The smallest upper bound on the $\mathbf{L}_2$ gain, provable using Lur'e Lyapunov functions, is therefore obtained by minimizing γ over γ, P, Λ and T subject to $P > 0$, $\Lambda = \mathbf{diag}(\lambda_1, \ldots, \lambda_{n_p}) \geq 0$, $T = \mathbf{diag}(\tau_1, \ldots, \tau_{n_p}) \geq 0$ and (8.12). This is an EVP.

8.2 Integral Quadratic Constraints

In this section we consider an important variation on the NLDI, in which the pointwise constraint $p(t)^Tp(t) \leq q(t)^Tq(t)$ is replaced with a constraint on the integrals of $p(t)^Tp(t)$ and $q(t)^Tq(t)$:

$$\begin{aligned} \dot{x} &= Ax + B_pp + B_uu + B_ww, \\ q &= C_qx + D_{qp}p + D_{qu}u + D_{qw}w, \\ z &= C_zx + D_{zp}p + D_{zu}u + D_{zw}w \\ \int_0^t p(\tau)^Tp(\tau)\,d\tau &\leq \int_0^t q(\tau)^Tq(\tau)\,d\tau. \end{aligned} \tag{8.13}$$

Such a system is closely related to the NLDI with the same data matrices; indeed, every trajectory of the associated NLDI is a trajectory of (8.13). This system is not, however, a differential inclusion. In control theory terms, the system (8.13) is described as a linear system with (dynamic) nonexpansive feedback. An important example is when p and q are related by a linear system, i.e.,

$$\dot{x}_f = A_f x_f + B_f q, \qquad p = C_f x_f + D_f q, \qquad x_f(0) = 0,$$

where $\|D_f + C_f(sI - A_f)^{-1}B_f\|_\infty \le 1$.

We can also consider a generalization of the DNLDI, in which we have componentwise integral quadratic constraints. Moreover, we can consider integral quadratic constraints with different sector bounds; as an example we will encounter constraints of the form $\int_0^t p(\tau)^T q(\tau)\, d\tau \ge 0$ in §8.3.

We will now show that many of the results on NLDIs (and DNLDIs) from Chapters 5–7 generalize to systems with integral quadratic constraints. We will demonstrate this generalization for stability analysis and $\mathbf{L}_2$ gain bounds, leaving others to the reader.

As in Chapters 5–7, our analysis will be based on quadratic functions of the state $V(\xi) = \xi^T P \xi$, and roughly speaking, the very same LMIs will arise. However, the interpretation of V is different here. For example, in stability analysis, the V of Chapters 5–7 decreases monotonically to zero. Here, the very same V decreases to zero, but not necessarily monotonically. Thus, V is not a Lyapunov function in the conventional sense.

8.2.1 Stability

Consider system (8.13) without w and z. Suppose $P > 0$ and $\lambda \ge 0$ are such that

$$\frac{d}{dt} x^T P x < \lambda \left(p^T p - q^T q\right), \tag{8.14}$$

or

$$\frac{d}{dt}\left(x^T P x + \lambda \int_0^t \left(q(\tau)^T q(\tau) - p(\tau)^T p(\tau)\right) d\tau\right) < 0.$$

Note that the second term is always nonnegative. Using standard arguments from Lyapunov theory, it can be shown that $\lim_{t\to\infty} x(t) = 0$, or the system is stable.

Now, let us examine the condition (8.14). It is exactly the same as the condition obtained by applying the $\mathcal{S}$-procedure to the condition

$$\frac{d}{dt} x^T P x < 0, \text{ whenever } p^T p \le q^T q,$$

which in turn, leads to the LMI condition for quadratic stability of NLDI (5.3)

$$P > 0, \qquad \lambda \ge 0,$$

$$\begin{bmatrix} A^T P + PA + \lambda C_q^T C_q & PB_p + \lambda C_q^T D_{qp} \\ (PB_p + \lambda C_q^T D_{qp})^T & -\lambda(I - D_{qp}^T D_{qp}) \end{bmatrix} < 0.$$

8.2.2 $\mathbf{L}_2$ gain

We assume that $x(0) = 0$ and, for simplicity, that $D_{zw} = 0$ and $D_{qp} = 0$.

Suppose $P > 0$ and $\lambda \geq 0$ are such that

$$\frac{d}{dt} x^T P x < \lambda \left(p^T p - q^T q \right) + \gamma^2 w^T w - z^T z. \tag{8.15}$$

Integrating both sides from 0 to T,

$$x(T)^T P x(T) + \lambda \int_0^T \left(q(\tau)^T q(\tau) - p(\tau)^T p(\tau) \right) d\tau < \int_0^T \left(\gamma^2 w^T w - z^T z \right) dt.$$

This implies that the $\mathbf{L}_2$ gain from w to z for system (8.13) does not exceed γ. Condition (8.15) leads to the same LMI (6.55) which guarantees that the $\mathbf{L}_2$ gain of the NLDI (4.9) from w to z does not exceed γ.

Remark: Almost all the results described in Chapters 5, 6 and 7 extend immediately to systems with integral quadratic constraints. The only exceptions are the results on coordinate transformations (§5.1.1, also §7.2.2), and on reachable sets with unit-peak inputs (§6.1.3 and §7.3.3).

8.3 Multipliers for Systems with Unknown Parameters

We consider the system

$$\begin{array}{ll} \dot{x} = Ax + B_p p + B_w w, & q = C_q x + D_{qp} p, \\ p_i = \delta_i q_i, \quad i = 1, \ldots, n_p, & z = C_z x + D_{zw} w, \end{array} \tag{8.16}$$

where $p(t) \in \mathbf{R}^{n_p}$, and δ_i, $i = 1, \ldots, n_p$ are any nonnegative numbers. We can consider the more general case when $\alpha_i \leq \delta_i \leq \beta_i$ using a loop transformation (see the Notes and References).

For reasons that will become clear shortly, we begin by defining n_p LTI systems with input q_i and output $q_{m,i}$, where $q_{m,i}(t) \in \mathbf{R}$:

$$\begin{array}{rcl} \dot{x}_{m,i} & = & A_{m,i} x_{m,i} + B_{m,i} q_i, \quad x_{m,i}(0) = 0, \\ q_{m,i} & = & C_{m,i} x_{m,i} + D_{m,i} q_i, \end{array} \tag{8.17}$$

where we assume that $A_{m,i}$ is stable and $(A_{m,i}, B_{m,i})$ is controllable. We further assume that

$$\int_0^t q_i(\tau) q_{m,i}(\tau) \, d\tau \geq 0 \text{ for all } t \geq 0.$$

This last condition is equivalent to passivity of the LTI systems (8.17), which in turn is equivalent to the existence of $P_i \geq 0$, $i = 1, \ldots, n_p$, satisfying

$$\left[\begin{array}{cc} A_{\mathrm{m},i}^T P_i + P_i A_{\mathrm{m},i} & P_i B_{\mathrm{m},i} - C_{\mathrm{m},i}^T \\ B_{\mathrm{m},i}^T P_i - C_{\mathrm{m},i} & -\left(D_{\mathrm{m},i} + D_{\mathrm{m},i}^T \right) \end{array} \right] \leq 0, \quad i = 1, \ldots, n_p. \tag{8.18}$$

(See §6.3.3.)

Thus, q and $q_m \stackrel{\Delta}{=} [q_{m,1}, \ldots, q_{m,n_p}]^T$ are the input and output respectively of the diagonal system with realization

$$A_m = \mathbf{diag}(A_{m,i}), \quad B_m = \mathbf{diag}(B_{m,i}), \quad C_m = \mathbf{diag}(C_{m,i}), \quad D_m = \mathbf{diag}(D_{m,i}),$$

Defining

$$\tilde{A} = \begin{bmatrix} A & 0 \\ B_m C_q & A_m \end{bmatrix}, \qquad \tilde{B}_p = \begin{bmatrix} B_p \\ B_m D_{qp} \end{bmatrix}, \qquad \tilde{B}_w = \begin{bmatrix} B_w \\ 0 \end{bmatrix},$$
$$\tilde{C}_q = [D_m C_q \;\; C_m], \qquad \tilde{D}_{qp} = D_m D_{qp}, \qquad \tilde{C}_z = [C_z \;\; 0].$$

consider the following system with integral quadratic constraints:

$$\begin{aligned} \frac{d\tilde{x}}{dt} &= \tilde{A}\tilde{x} + \tilde{B}_p p + \tilde{B}_w w, \\ q_m &= \tilde{C}_q \tilde{x} + \tilde{D}_{qp} p, \\ z &= \tilde{C}_z \tilde{x} + \tilde{D}_{qp} p + D_{zw} w, \\ &\int_0^t p_i(\tau) q_{m,i}(\tau)\, d\tau \geq 0, \quad i = 1, \ldots, n_p. \end{aligned} \tag{8.19}$$

It is easy to show that if x is a trajectory of (8.16), then $[x^T \;\; x_m^T]^T$ is a trajectory of (8.19) for some appropriate x_m. Therefore, conclusions about (8.16) have implications for (8.19). For instance, if system (8.19) is stable, so is system (8.16).

8.3.1 Stability

Consider system (8.19) without w and z. Suppose $P > 0$ is such that

$$\frac{d}{dt}\tilde{x}^T P \tilde{x} + 2\sum_{i=1}^{n_p} p_i q_{m,i} < 0, \tag{8.20}$$

or

$$\frac{d}{dt}\left(\tilde{x}^T P \tilde{x} + 2\sum_{i=1}^{n_p} \int_0^t p_i(\tau) q_{m,i}(\tau)\, d\tau\right) < 0.$$

Then, it can be shown using standard arguments that $\lim_{t\to\infty} \tilde{x}(t) = 0$, i.e., the system (8.19) is stable.

Condition (8.20) is just the LMI in $P > 0$, C_m and D_m:

$$\begin{bmatrix} \tilde{A}^T P + P\tilde{A} & P\tilde{B}_p + \tilde{C}_q^T \\ \tilde{B}_p^T P + \tilde{C}_q & \tilde{D}_{qp} + \tilde{D}_{qp}^T \end{bmatrix} < 0. \tag{8.21}$$

Remark: LMI (8.21) can also be interpreted as a positive dissipation condition for the LTI system with transfer matrix $G(s) = -W(s)H(s)$, where $W(s) = C_m(sI - A_m)^{-1}B_m + D_m$ and $H(s) = C_q(sI - A)^{-1}B_p + D_{qp}$. Therefore, we conclude that system (8.16) is stable if there exists a passive W such that $-W(s)H(s)$ has positive dissipation. Hence W is often called a "multiplier". See the Notes and References for details.

8.3.2 Reachable sets with unit-energy inputs

The set of reachable states with unit-energy inputs for the system

$$\begin{aligned} \dot{x} &= Ax + B_p p + B_w w, \qquad q = C_q x + D_{qp} p, \\ p &= \delta q, \qquad \delta \geq 0, \end{aligned} \tag{8.22}$$

is

$$\mathcal{R}_{\rm ue} \triangleq \left\{ x(T) \;\middle|\; \begin{array}{ll} x,\, w \text{ satisfy (8.22)}, & x(0)=0 \\ \int_0^T w^T w \, dt \le 1, & T \ge 0 \end{array} \right\}.$$

We can obtain a sufficient condition for a point x_0 to lie outside the reachable set $\mathcal{R}_{\rm ue}$, using quadratic functions for the augmented system. Suppose $P > 0$ is such that

$$\frac{d}{dt}\tilde{x}^T P \tilde{x} + 2\sum_{i=1}^{n_p} p_i q_{m,i} < w^T w, \tag{8.23}$$

or

$$\tilde{x}^T P \tilde{x} + 2\sum_{i=1}^{n_p} \int_0^t p_i(\tau) q_{m,i}(\tau)\, d\tau < \int_0^t w(\tau)^T w(\tau)\, d\tau.$$

then the set $\mathcal{E} = \left\{ \tilde{x} \mid \tilde{x}^T P \tilde{x} \le 1 \right\}$ contains the reachable set for the system (8.19). Condition (8.23) is equivalent to the LMI in $P > 0$, C_m and D_m:

$$\begin{bmatrix} \tilde{A}^T P + P\tilde{A} & P\tilde{B}_w & P\tilde{B}_p + \tilde{C}_q^T \\ \tilde{B}_w^T P & -I & 0 \\ \tilde{B}_p^T P + \tilde{C}_q & 0 & \tilde{D}_{qp} + \tilde{D}_{qp}^T \end{bmatrix} < 0. \tag{8.24}$$

Now, the point x_0 lies outside the set of reachable states for the system (8.22) if there exists P satisfying (8.23) and there exists no z such that

$$\begin{bmatrix} x_0 \\ z \end{bmatrix}^T P \begin{bmatrix} x_0 \\ z \end{bmatrix} \le 1.$$

Partitioning P conformally with the sizes of x_0 and z as

$$P = \begin{bmatrix} P_{11} & P_{12} \\ P_{12}^T & P_{22} \end{bmatrix},$$

we note that

$$\min_z \begin{bmatrix} x_0 \\ z \end{bmatrix}^T \begin{bmatrix} P_{11} & P_{12} \\ P_{12}^T & P_{22} \end{bmatrix} \begin{bmatrix} x_0 \\ z \end{bmatrix} = x_0^T (P_{11} - P_{12} P_{22}^{-1} P_{12}^T) x_0.$$

Therefore, x_0 does not belong to the reachable set for the system (8.22) if there exist $P > 0$, $P_i \ge 0$, $\lambda \ge 0$, C_m, and $D_{\rm M}$ satisfying (8.18), (8.24) and

$$\begin{bmatrix} x_0^T P_{11} x_0 - 1 & x_0^T P_{12} \\ P_{12}^T x_0 & P_{22} \end{bmatrix} > 0.$$

This is an LMIP.

Notes and References

Lur'e systems

The Notes and References of Chapters 5 and 7 are relevant to this chapter as well. Lur'e and Postnikov were the first to propose Lyapunov functions consisting of a quadratic form plus the integral of the nonlinearity [LP44]. The book [LUR57] followed this article, in which the author showed how to solve the problem of finding such Lyapunov functions analytically for systems of order 2 or 3. Another early book that uses such Lyapunov functions is Letov [LET61]. Work on this topic was continued by Popov [POP62], Yakubovich [YAK64]. Tsypkin [TSY64D, TSY64C, TSY64A, TSY64B], Szego [SZE63] and Meyer [MEY66]. Jury and Lee [JL65] consider the case of Lur'e systems with multiple nonlinearities and derive a corresponding stability criterion. See also [IR64].

The ellipsoid method is used to construct Lyapunov functions for the Lur'e stability problem in the papers by Pyatnitskii and Skorodinskii [PS82, PS83]. See also [SKO91A, SKO91B] for general numerical methods to prove stability of nonlinear systems via the Popov or circle criterion. A precursor is the paper by Karmarkar and Siljak [KS75], who used a locally convergent algorithm to determine margins for a Lur'e-Postnikov system with one nonlinearity.

Indeed, in the early books by Lur'e, Postnikov, and Letov the problem of constructing Lyapunov functions reduces to the satisfaction of some inequalities. In some special cases they even point out geometrical properties of the regions defined by the inequalities, i.e., that they form a parallelogram. But as far as we know, convexity of the regions is not noted in the early work. In any case, it was not known in the 1940's and 1950's that solution of convex inequalities is practically and theoretically tractable.

Construction of Lyapunov functions for systems with integral quadratic constraints

We saw in §8.2.1 that the quadratic positive function $V(\xi) = \xi^T P\xi$ is not necessarily a Lyapunov function in the conventional sense for the system (8.13) with integral quadratic constraints; although $V(x(t)) \to 0$ as $t \to \infty$, it does not do so monotonically. We now show how we can explicitly construct a Lyapunov function given P, in cases when p and q are related by

$$\dot{x}_f = f(x_f, q, t), \qquad p = g(x_f, q, t), \tag{8.25}$$

where $x_f(t) \in \mathbf{R}^{n_f}$.

Since $\int_0^t p^T p\, d\tau \leq \int_0^t q^T q\, d\tau$, it can be shown (see for example [WIL72]) that $V_f : \mathbf{R}^{n_f} \to \mathbf{R}_+$ given by

$$V_f(\xi_f) \stackrel{\Delta}{=} \sup \left\{ -\int_0^T \left(q^T q - p^T p\right) \, dt \;\middle|\; x_f(0) = \xi_f,\ x_f \text{ satisfies (8.25)},\ T \geq 0 \right\}$$

is a Lyapunov function that proves stability of system (8.25). Moreover, the Lyapunov function $V : \mathbf{R}^n \times \mathbf{R}^{n_f} \to \mathbf{R}_+$ given by

$$V(\xi, \xi_f) \stackrel{\Delta}{=} \xi^T P\xi + V_f(\xi_f)$$

proves stability of the interconnection

$$\begin{aligned} \dot{x} &= Ax + B_p p, & q &= C_q x + D_{qp} p, \\ \dot{x}_f &= f(x_f, q, t), & p &= g(x_f, q, t). \end{aligned}$$

Note that this construction also shows that $\xi^T P\xi$ proves the stability of the NLDI (5.2).

Integral quadratic constraints have recently received much attention, especially in the former Soviet Union. See for example [YAK88, MEG92A, MEG92B, YAK92, SAV91, SP94B, SP94H, SP94D].

Systems with constant, unknown parameters

Problems involving system (8.16) are widely studied in the field of robust control, sometimes under the name "real-μ problems"; see Siljak [SIL89] for a survey. The results of §8.3 were derived by Fan, Tits and Doyle in [FTD91] using an alternate approach, involving the application of an off-axis circle criterion (see also §3.3). We also mention [TV91], where Tesi and Vicino study a hybrid system, consisting of a Lur'e system with unknown parameters.

There are elegant analytic solutions for some very special problems involving parameter-dependent linear systems, e.g., Kharitonov's theorem [KHA78] and its extensions [BHL89]. A nice survey of these results can be found in Barmish [BAR93].

Complexity of stabilization problems

The stability (and performance) analysis problems considered in §8.3 have high complexity. For example, checking if (8.16) is stable is NP-hard (see Coxson and Demarco [CD91], Braatz and Young [BYDM93], Poljak and Rohn [PR94], and Nemirovskii [NEM94]). We conjecture that checking whether a general DNLDI is stable is also NP-hard.

Likewise, many stabilization problems for uncertain systems are NP-hard. For instance, the problem of checking whether system (8.16) is stabilizable by a constant state-feedback can be shown to be NP-hard using the method of [NEM94].

A related result, due to Blondel and Gevers [BG94], states that checking whether there exists a common stabilizing LTI controller for three LTI systems is undecidable. In contrast, checking whether there exists a static, output-feedback control law for a *single* LTI system is rationally decidable [ABJ75].

Multiplier methods

See [WIL69A, NT73, DV75, SIL69, WIL70, WIL76] for discussion of and bibliography on multiplier theory and its connections to Lyapunov stability theory. In general multiplier theory we consider the system

$$\dot{x} = Ax + B_p p + B_w w, \qquad q = C_q x + D_{qp} p, \qquad p(t) = \Delta(q, t).$$

The method involves checking that $H_m = W\left(C_q(sI - A)^{-1}B_p + D_{qp}\right)$ satisfies

$$H_m(j\omega) + H_m(j\omega)^* \geq 0 \text{ for every } \omega \in \mathbf{R},$$

for some choice of W from a set that is determined by the properties of Δ. When Δ is a nonlinear time-invariant memoryless operator satisfying a sector condition (i.e., when we have a Lur'e system), the multipliers W are of the form $(1 + qs)$ for some $q \geq 0$. This is the famous Popov criterion [POP62, POP64]; we must also cite Yakubovich, who showed, using the $\mathcal{S}$-procedure [YAK77], that feasibility of LMI (8.6) is equivalent to the existence of q nonnegative such that $(1 + qs)(c_q(sI - A)^{-1}b_p) + 1/k$ has positive dissipation.

In the case when $\Delta(q, t) = \delta q(t)$ for some unknown real number δ, the multiplier W can be any passive transfer matrix. This observation was made by Brockett and Willems in [BW65]; given a transfer function H_{qp} they show that the transfer function $H_{qp}/(1+kH_{qp})$ is stable for all values of k in $(0, \infty)$ if and only if there exists a passive multiplier W such that WH_{qp} is also passive. This case has also been considered in [CS92B, SC93, SL93A, SLC94] where the authors devise appropriate multipliers for constant real uncertainties. The paper [BHPD94] by Balakrishnan et al., shows how various stability tests for uncertain systems can be rederived in the context of multiplier theory, and how these tests can be reduced to LMIPs. See also [LSC94].

Safonov and Wyetzner use a convex parametrization of multipliers found by Zames and Falb [ZF68, ZF67] to prove stability of systems subject to "monotonic" or "odd-monotonic" nonlinearities (see [SW87] for details; see also [GG94]). Hall and How, along with Bernstein and Haddad, generalize the use of quadratic Lyapunov functions along with multipliers for various classes of nonlinearities, and apply them to Aerospace problems; see for example [HB91A, HOW93, HH93B, HHHB92, HH93A, HHH93A, HCB93, HB93B, HB93A, HHH93B]. See also [CHD93].

Multiple nonlinearities for Lur'e systems

Our analysis method for Lur'e systems with multiple nonlinearities in §8.1 can be conservative, since the $\mathcal{S}$-procedure can be conservative in this case. Rapoport [RAP86, RAP87, RAP88] devises a way of determining if there exists Lyapunov functions of the form (8.3) nonconservatively by generating appropriate LMIs. We note, however, that the number of these LMIs grows exponentially with the number of nonlinearities. Kamenetskii [KAM89] and Rapoport [RAP89] derive corresponding frequency-domain stability criteria. Kamenetskii calls his results "convolution method for solving matrix inequalities".

Loop transformations

Consider the Lur'e system with general sector conditions, i.e.,

$$\begin{array}{ll} \dot{x} = Ax + Bp, & q = Cx + Dp, \\ p_i(t) = \phi_i(q_i(t)), & \alpha_i \sigma^2 \le \sigma\phi_i(\sigma) \le \beta_i\sigma^2. \end{array} \tag{8.26}$$

Here we have dropped the variables w and z and the subscripts on B, C and D to simplify the presentation. We assume this system is well-posed, i.e., $\det(I - D\Delta) \neq 0$ for all diagonal Δ with $\alpha_i \le \Delta_{ii} \le \beta_i$.

Define

$$\bar{p}_i \stackrel{\Delta}{=} \frac{1}{\beta_i - \alpha_i}\left(\phi_i(q_i) - \alpha_i q_i\right) = \bar{\phi}_i(q_i).$$

It is readily shown that $0 \le \sigma\bar{\phi}_i(\sigma) \le \sigma^2$ for all σ. Let Λ and Γ denote the diagonal matrices

$$\Lambda = \mathbf{diag}(\alpha_1, \dots, \alpha_{n_q}), \qquad \Gamma = \mathbf{diag}(\beta_1 - \alpha_1, \dots, \beta_{n_q} - \alpha_{n_q}),$$

so that $\bar{p} = \Gamma^{-1}(p - \Lambda q)$. We now substitute $p = \Gamma\bar{p} + \Lambda q$ into (8.26) and, using our well-posedness assumption, solve for $\dot{x}$ and q in terms of x and $\bar{p}$. This results in

$$\begin{array}{rcl} \dot{x} &=& \left(A + B\Lambda(I - D\Lambda)^{-1}C\right)x + B(I - \Lambda D)^{-1}\Gamma\bar{p}, \\ q &=& (I - D\Lambda)^{-1}Cx + (I - D\Lambda)^{-1}D\Gamma\bar{p}. \end{array}$$

We can therefore express (8.26) as

$$\begin{array}{ll} \dot{x} = \bar{A}x + \bar{B}\bar{p}, & q = \bar{C}x + \bar{D}\bar{p}, \\ \bar{p}_i(t) = \bar{\phi}_i(q_i(t)), & 0 \le \sigma\bar{\phi}_i(\sigma) \le \sigma^2. \end{array} \tag{8.27}$$

where

$$\begin{array}{ll} \bar{A} = A + B\Lambda(I - D\Lambda)^{-1}C, & \bar{B} = B(I - \Lambda D)^{-1}\Gamma, \\ \bar{C} = (I - D\Lambda)^{-1}C, & \bar{D} = (I - D\Lambda)^{-1}D\Gamma. \end{array}$$

Note that (8.27) is in the standard Lur'e system form. So to analyze the more general Lur'e system (8.26), we simply apply the methods of §8.1 to the loop-transformed system (8.27). In practice, i.e., in a numerical implementation, it is probably better to derive the LMIs associated with the more general Lur'e system than to loop transform to the standard Lur'e system.

The construction above is a loop transformation that maps nonlinearities in sector $[\alpha_i, \beta_i]$ into the standard sector, i.e., $[0, 1]$. Similar transformations can be used to map any set of sectors into any other, including so-called infinite sectors, in which, say, $\beta_i = +\infty$.

The term loop transformation comes from a simple block-diagram interpretation of the equations given above. For detailed descriptions of loop transformations, see the book by Desoer and Vidyasagar [DV75, P50].

Chapter 9

Systems with Multiplicative Noise

9.1 Analysis of Systems with Multiplicative Noise

9.1.1 Mean-square stability

We first consider the discrete-time stochastic system

$$x(k+1) = \left(A_0 + \sum_{i=1}^{L} A_i p_i(k) \right) x(k), \tag{9.1}$$

where $p(0)$, $p(1)$, ..., are independent, identically distributed random variables with

$$\mathbf{E}\, p(k) = 0, \qquad \mathbf{E}\, p(k)p(k)^T = \Sigma = \mathbf{diag}(\sigma_1, \dots, \sigma_L). \tag{9.2}$$

We assume that $x(0)$ is independent of the process p.

Define $M(k)$, the state correlation matrix, as

$$M(k) \triangleq \mathbf{E}\, x(k)x(k)^T.$$

Of course, M satisfies the linear recursion

$$M(k+1) = AM(k)A^T + \sum_{i=1}^{L} \sigma_i^2 A_i M(k) A_i^T, \quad M(0) = \mathbf{E}\, x(0)x(0)^T. \tag{9.3}$$

If this linear recursion is stable, i.e., regardless of $x(0)$, $\lim_{k\to\infty} M(k) = 0$, we say the system is *mean-square stable*. Mean-square stability implies, for example, that $x(k) \to 0$ almost surely.

It can be shown (see the Notes and References) that mean-square stability is equivalent to the existence of a matrix $P > 0$ satisfying the LMI

$$A^T P A - P + \sum_{i=1}^{L} \sigma_i^2 A_i^T P A_i < 0. \tag{9.4}$$

An alternative, equivalent condition is that the dual inequality

$$AQA^T - Q + \sum_{i=1}^{L} \sigma_i^2 A_i Q A_i^T < 0 \tag{9.5}$$

holds for some $Q > 0$.

Remark: For $P > 0$ satisfying (9.4), we can interpret the function $V(\xi) = \xi^T P\xi$ as a *stochastic Lyapunov function*: $\mathbf{E}\, V(x)$ decreases along trajectories of (9.1). Alternatively, the function $V(M) = \mathbf{Tr}\, MP$ is a (linear) Lyapunov function for the deterministic system (9.3) (see the Notes and References).

Remark: Mean-square stability can be verified directly by solving the Lyapunov equation in P

$$A^T PA - P + \sum_{i=1}^{L} \sigma_i^2 A_i^T PA_i + I = 0$$

and checking whether $P > 0$. Thus the LMIPs (9.4) and (9.5) offer no computational (or theoretical) advantages for checking mean-square stability.

We can consider variations on the mean-square stability problem, for example, determining the mean-square stability margin. Here, we are given $A_0, \ldots, A_L$, and asked to find the largest γ such that with $\Sigma < \gamma^2 I$, the system (9.1) is mean-square stable. From the LMIP characterizations of mean-square stability, we can derive GEVP characterizations that yield the exact mean-square stability margin. Again, the GEVP offers no computational or theoretical advantages, since the mean-square stability margin can be obtained by computing the eigenvalue of a (large) matrix.

9.1.2 State mean and covariance bounds with unit-energy inputs

We now add an exogenous input w to our stochastic system:

$$x(k+1) = Ax(k) + B_w w(k) + \sum_{i=1}^{L} \left(A_i x(k) + B_{w,i} w(k)\right) p_i(k), \quad x(0) = 0. \quad (9.6)$$

We assume that the exogenous input w is deterministic, with energy not exceeding one, i.e.,

$$\sum_{k=0}^{\infty} w(k)^T w(k) \le 1.$$

Let $\bar{x}(k)$ denote the mean of $x(k)$, which satisfies $\bar{x}(k+1) = A\bar{x}(k) + B_w w(k)$, and let $X(k)$ denote the covariance of $x(k)$, i.e., $X(k) = \mathbf{E}\left(x(k) - \bar{x}(k)\right)\left(x(k) - \bar{x}(k)\right)^T$. We will develop joint bounds on $\bar{x}(k)$ and $X(k)$.

Suppose $V(\xi) = \xi^T P\xi$ where $P > 0$ satisfies

$$\mathbf{E}\, V(x(k+1)) \le \mathbf{E}\, V(x(k)) + w(k)^T w(k) \quad (9.7)$$

for all trajectories. Then,

$$\mathbf{E}\, V(x(k)) \le \sum_{j=0}^{k} w(j)^T w(j), \quad k \ge 0,$$

which implies that, for all $k \ge 0$,

$$\begin{aligned} \mathbf{E}\, x(k)^T P x(k) &= \bar{x}(k)^T P\bar{x}(k) + \mathbf{E}\left(x(k) - \bar{x}(k)\right)^T P\left(x(k) - \bar{x}(k)\right) \\ &= \bar{x}(k)^T P\bar{x}(k) + \mathbf{Tr}\, X(k)P \\ &\le 1. \end{aligned}$$

Let us now examine condition (9.7). We note that

$$\begin{aligned} \mathbf{E}\, V(x(k+1)) &= \begin{bmatrix} \bar{x}(k) \\ w(k) \end{bmatrix}^T M \begin{bmatrix} \bar{x}(k) \\ w(k) \end{bmatrix} \\ &\quad + \mathbf{Tr} \left(A^T P A + \sum_{i=1}^{L} \sigma_i^2 A_i^T P A_i \right) X(k), \end{aligned}$$

where

$$M = \begin{bmatrix} A^T \\ B_w^T \end{bmatrix} P \begin{bmatrix} A & B_w \end{bmatrix} + \sum_{i=1}^{L} \sigma_i^2 \begin{bmatrix} A_i^T \\ B_{w,i}^T \end{bmatrix} P \begin{bmatrix} A_i & B_{w,i} \end{bmatrix}.$$

Next, the LMI conditions

$$\begin{bmatrix} A^T \\ B_w^T \end{bmatrix} P \begin{bmatrix} A & B_w \end{bmatrix} + \sum_{i=1}^{L} \sigma_i^2 \begin{bmatrix} A_i^T \\ B_{w,i}^T \end{bmatrix} P \begin{bmatrix} A_i & B_{w,i} \end{bmatrix} \leq \begin{bmatrix} P & 0 \\ 0 & I \end{bmatrix} \tag{9.8}$$

and

$$A^T P A + \sum_{i=1}^{L} \sigma_i^2 A_i^T P A_i < P \tag{9.9}$$

imply

$$\begin{aligned} \mathbf{E}\, V(x(k+1)) &\leq \bar{x}^T(k) P \bar{x}(k) + w(k)^T w(k) \\ &\quad + \mathbf{Tr} \left(A^T P A + \sum_{i=1}^{L} \sigma_i^2 A_i^T P A_i \right) X(k) \\ &\leq \bar{x}^T(k) P \bar{x}(k) + w(k)^T w(k) \\ &\quad + \mathbf{E} \left(x(k) - \bar{x}(k) \right)^T P \left(x(k) - \bar{x}(k) \right) \\ &= \mathbf{E}\, V(x(k)) + w(k)^T w(k), \quad k \geq 0. \end{aligned}$$

Of course, LMI (9.8) implies LMI (9.9); thus, we conclude that LMI (9.8) implies (9.7). Therefore, we conclude that $\bar{x}(k)$ and $X(k)$ must belong to the set

$$\left\{ (\bar{x(k)}, X(k)) \;\middle|\; \bar{x}(k)^T P \bar{x}(k) + \mathbf{Tr}\, X(k) P \leq 1 \right\}$$

for any $P > 0$ satisfying the LMI (9.8). For example, we have the bound $\mathbf{Tr}\, M(k) P \leq 1$ on the state covariance matrix.

We can derive further bounds on $M(k)$. For example, since we have $\mathbf{Tr}\, M(k) P \leq \lambda_{\min}(P)\, \mathbf{Tr}\, M(k)$ and therefore $\mathbf{Tr}\, M(k) \leq 1/\lambda_{\min}(P)$, we can derive bounds on $\mathbf{Tr}\, M$ by maximizing $\lambda_{\min}(P)$ subject to (9.8) and $P > 0$. This is an EVP. As another example, we can derive an upper bound on $\lambda_{\max}(M)$ by maximizing $\mathbf{Tr}\, P$ subject to (9.8) and $P > 0$, which is another EVP.

9.1.3 Bound on $\mathbf{L_2}$ gain

We now consider the system

$$\begin{aligned} x(k+1) &= Ax(k) + B_w w(k) + \sum_{i=1}^{L} \left(A_i x(k) + B_{w,i} w(k) \right) p_i(k), \quad x(0) = 0, \\ z(k) &= C_z x(k) + D_{zw} w(k) + \sum_{i=1}^{L} \left(C_{z,i} x(k) + D_{zw,i} w(k) \right) p_i(k). \end{aligned} \tag{9.10}$$

with the same assumptions on p. We assume that w is deterministic.

We define the $\mathbf{L}_2$ gain η of this system as

$$\eta^2 \triangleq \sup \left\{ \mathbf{E} \sum_{k=0}^{\infty} z(k)^T z(k) \;\middle|\; \sum_{k=0}^{\infty} w(k)^T w(k) \le 1 \right\}. \tag{9.11}$$

Suppose that $V(\xi) = \xi^T P \xi$, with $P > 0$, satisfies

$$\mathbf{E}\, V(x(k+1)) - \mathbf{E}\, V(x(k)) \le \gamma^2 w(k)^T w(k) - \mathbf{E}\, z(k)^T z(k). \tag{9.12}$$

Then $\gamma \ge \eta$. The condition (9.12) can be shown to be equivalent to the LMI

$$\begin{array}{l} \begin{bmatrix} A & B_w \\ C_z & D_{zw} \end{bmatrix}^T \begin{bmatrix} P & 0 \\ 0 & I \end{bmatrix} \begin{bmatrix} A & B_w \\ C_z & D_{zw} \end{bmatrix} - \begin{bmatrix} P & 0 \\ 0 & \gamma^2 I \end{bmatrix} \\ \quad + \displaystyle\sum_{i=1}^{L} \sigma_i^2 \begin{bmatrix} A_i & B_{w,i} \\ C_{z,i} & D_{zw,i} \end{bmatrix}^T \begin{bmatrix} P & 0 \\ 0 & I \end{bmatrix} \begin{bmatrix} A_i & B_{w,i} \\ C_{z,i} & D_{zw,i} \end{bmatrix} \le 0. \end{array} \tag{9.13}$$

Minimizing γ subject to LMI (9.13), which is an EVP, yields an upper bound on η.

Remark: It is straightforward to apply the scaling method developed in §6.3.4 to obtain componentwise results.

9.2 State-Feedback Synthesis

We now add a control input u to our system:

$$\begin{array}{rcl} x(k+1) & = & Ax(k) + B_u u(k) + B_w w(k) + \\ & & \displaystyle\sum_{i=1}^{L} \left(A_i x(k) + B_{u,i} u(k) + B_{w,i} w(k)\right) p_i(k), \\ z(k) & = & C_z x(k) + D_{zu} u(k) + D_{zw} w(k) + \\ & & \displaystyle\sum_{i=1}^{L} \left(C_{z,i} x(k) + D_{zu,i} u(k) + D_{zw,i} w(k)\right) p_i(k). \end{array} \tag{9.14}$$

We seek a state-feedback $u(k) = Kx(k)$ such that the closed-loop system

$$\begin{array}{rcl} x(k+1) & = & (A + B_u K) x(k) + B_w w(k) + \\ & & \displaystyle\sum_{i=1}^{L} \left((A_i + B_{u,i} K) x(k) + B_{w,i} w(k)\right) p_i(k), \\ z(k) & = & (C_z + D_{zu} K) x(k) + D_{zw} w(k) + \\ & & \displaystyle\sum_{i=1}^{L} \left((C_{z,i} + D_{zu,i} K) x(k) + D_{zw,i} w(k)\right) p_i(k), \end{array} \tag{9.15}$$

satisfies various properties.

9.2.1 Stabilizability

We seek K and $Q > 0$ such that

$$\begin{array}{l} (A + B_u K) Q (A + B_u K)^T - Q \\ \qquad + \displaystyle\sum_{i=1}^{L} \sigma_i^2 (A_i + B_{u,i} K) Q (A_i + B_{u,i} K)^T < 0. \end{array}$$

With the change of variable $Y = KQ$, we obtain the equivalent condition

$$\begin{aligned}(AQ + B_uY)Q^{-1}(AQ + B_uY)^T - Q \\ + \sum_{i=1}^{L} \sigma_i^2 (A_iQ + B_{u,i}Y)Q^{-1}(A_iQ + B_{u,i}Y)^T < 0,\end{aligned} \tag{9.16}$$

which is easily written as an LMI in Q and Y.

Thus, system (9.14) is mean-square stabilizable (with constant state-feedback) if and only if the LMI $Q > 0$, (9.16) is feasible. In this case, a stabilizing state-feedback gain is $K = YQ^{-1}$.

As an extension, we can maximize the closed-loop mean-square stability margin. System (9.15) is mean-square stable for $\Sigma \leq \gamma^2 I$ if and only if

$$\begin{aligned}(AQ + B_uY)Q^{-1}(AQ + B_uY)^T - Q \\ + \gamma^2 \sum_{i=1}^{L} (A_iQ + B_{u,i}Y)Q^{-1}(A_iQ + B_{u,i}Y)^T < 0\end{aligned} \tag{9.17}$$

holds for some Y and $Q > 0$. Therefore, finding K that maximizes γ reduces to the following GEVP:

$$\begin{array}{ll} \text{maximize} & \gamma \\ \text{subject to} & Q > 0, \quad (9.17) \end{array}$$

9.2.2 Minimizing the bound on $\mathbf{L}_2$ gain

We now seek a state-feedback gain K which minimizes the bound on $\mathbf{L}_2$ gain (as defined by (9.11)) for system (9.15). $\eta \leq \gamma$ for some K if there exist $P > 0$ and K such that

$$\begin{aligned}&\sum_{i=1}^{L} \sigma_i^2 \begin{bmatrix} A_i + B_{u,i}K & B_{w,i} \\ C_{z,i} + D_{zu,i}K & D_{zw,i} \end{bmatrix}^T \tilde{P} \begin{bmatrix} A_i + B_{u,i}K & B_{w,i} \\ C_{z,i} + D_{zu,i}K & D_{zw,i} \end{bmatrix} \\ &+ \begin{bmatrix} A + B_uK & B_w \\ C_z + D_{zu}K & D_{zw} \end{bmatrix}^T \tilde{P} \begin{bmatrix} A + B_uK & B_w \\ C_z + D_{zu}K & D_{zw} \end{bmatrix} \leq \begin{bmatrix} P & 0 \\ 0 & \gamma^2 I \end{bmatrix}\end{aligned}$$

where

$$\tilde{P} = \begin{bmatrix} P & 0 \\ 0 & I \end{bmatrix}.$$

We make the change of variables $Q = P^{-1}$, $Y = KQ$. Applying a congruence with $\tilde{Q} = \mathbf{diag}(Q, I)$, we get

$$\begin{aligned}&\sum_{i=1}^{L} \sigma_i^2 \begin{bmatrix} A_iQ + B_{u,i}Y & B_{w,i} \\ C_{z,i}Q + D_{zu,i}Y & D_{zw,i} \end{bmatrix}^T \tilde{Q}^{-1} \begin{bmatrix} A_iQ + B_{u,i}Y & B_{w,i} \\ C_{z,i}Q + D_{zu,i}Y & D_{zw,i} \end{bmatrix} \\ &+ \begin{bmatrix} AQ + B_uY & B_w \\ C_zQ + D_{zu}Y & D_{zw} \end{bmatrix}^T \tilde{Q}^{-1} \begin{bmatrix} AQ + B_uY & B_w \\ C_zQ + D_{zu}Y & D_{zw} \end{bmatrix} \leq \begin{bmatrix} Q & 0 \\ 0 & \gamma^2 I \end{bmatrix}.\end{aligned}$$

This inequality is readily transformed to an LMI.

Notes and References

Systems with multiplicative noise

Systems with multiplicative noise are a special case of *stochastic systems*. For an introduction to this topic see Arnold [ARN74] or Åström [ÅST70].

A rigorous treatment of continuous-time stochastic systems requires much technical machinery (see [EG94]). In fact there are many ways to formulate the continuous-time version of a system such as (9.10); see Itô [ITO51] and Stratonovitch [STR66].

Stability of system with multiplicative noise

For a survey of different concepts of stability for stochastic systems see [KOZ69] and the references therein. The definition of mean-square stability can be found in [SAM59]. Mean-square stability is a strong form of stability. In particular, it implies stability of the mean $\mathbf{E}\, x(k)$ and that all trajectories converge to zero with probability one [KUS67, WIL73]. For the systems considered here, it is also equivalent to $\mathbf{L}_2$-stability, which is the property that (see for example, [EP92]) $\mathbf{E} \sum_{t=0}^{T} \|x(t)\|^2$ has a (finite) limit as $T \to \infty$ for all initial conditions.

The equation for the state correlation matrix (9.3) can be found in [MCL71, WIL73]. An algebraic criterion for mean-square stability of a system with multiplicative white noise is given in [NS72]. It is reminiscent of the classical (deterministic) Hurwitz criterion. Frequency-domain criteria are given in [WB71]. The related Lyapunov equation was given by Kleinmann [KLE69] for the case $L = 1$ (i.e., a single uncertainty). The criterion then reduces to an $\mathbf{H}_2$ norm condition on a certain transfer matrix. (This is in parallel with the results of [EP92], see below.)

Kats and Krasovskii introduced in [KK60] the idea of a stochastic Lyapunov function and proved a related Lyapunov theorem. A thorough stability analysis of (nonlinear) stochastic systems was made by Kushner in [KUS67] using this idea of a stochastic Lyapunov function. Kushner used this idea to derive bounds for return time, expected output energy, etc. However, this approach does not provide a way to construct the Lyapunov functions.

Proof of stability criterion

We now prove that condition (9.4) is a necessary and sufficient condition for mean-square stability. We first prove that condition (9.4) is sufficient. Let $P > 0$ satisfy (9.4). Introduce the (linear Lyapunov) function $V(M) = \mathbf{Tr}\, MP$, which is positive on the cone of nonnegative matrices. For nonzero $M(k)$ satisfying (9.3),

$$\begin{aligned} V(M(k+1)) &= \mathbf{Tr} \left(AM(k)A^T + \sum_{i=1}^{L} \sigma_i^2 A_i M(k) A_i^T \right) P \\ &= \mathbf{Tr}\, M(k) \left(A^T P A + \sum_{i=1}^{L} \sigma_i^2 A_i^T P A_i \right) \\ &< \mathbf{Tr}\, M(k) P = V(M(k)). \end{aligned}$$

A further standard argument from Lyapunov theory proves that $M(k)$ converges to zero as $k \to \infty$.

To prove that condition (9.4) is also necessary, we assume that the system is mean-square stable. This means that for every positive initial condition $M(0) \geq 0$, the solution to the linear recursion (9.3) goes to zero as $k \to 0$. Linearity of system (9.3) implies that, for an arbitrary choice of $M(0)$, the corresponding solution $M(k)$ converges to 0, i.e., (9.3) is stable. We now write (9.3) as $m(k+1) = \mathcal{A} m(k)$, where $m(k)$ is a vector containing the n^2 elements of $M(k)$, and $\mathcal{A}$ is a real matrix. (This matrix can be expressed via Kronecker products involving $A_0, \ldots, A_L$.) Stability of $\mathcal{A}$ is equivalent to that of $\mathcal{A}^T$, and the corresponding matrix difference equation,

$$N(k+1) = A^T N(k) A + \sum_{i=1}^{L} \sigma_i^2 A_i^T N(k) A_i, \tag{9.18}$$

is also stable.

Now let $N(0) > 0$, and let $N(k)$ be the corresponding solution of (9.18). Since $N(k)$ satisfies a stable first-order linear difference equation with constant coefficients, the function

$$P(k) = \sum_{j=0}^{k} N(j)$$

has a (finite) limit for $k \to \infty$, which we denote by P. The fact that $N(0) > 0$ implies $P > 0$. Now (9.18) implies

$$P(k+1) = N(0) + A^T P(k) A + \sum_{i=1}^{L} \sigma_i^2 A_i^T P(k) A_i$$

Taking the limit $k \to \infty$ shows that P satisfies (9.4).

Mean-square stability margin

System (9.10) can be viewed as a linear system subject to random parameter variations. For systems with deterministic parameters only known to lie in ranges, the computation of an exact stability margin is an NP-hard problem [CD92]. The fact that a stochastic analog of the deterministic stability margin is easier to compute should not be surprising; we are dealing with a very special form of stability, which only considers the average behavior of the system.

An exact stability margin can also be computed for continuous-time systems with multiplicative noise provided the Itô framework is considered; see [EG94]. In contrast, the computation of an exact stability margin seems difficult for Stratonovitch systems.

In [EP92], El Bouhtouri and Pritchard provide a complete robustness analysis in a slightly different framework than ours, namely for so-called "block-diagonal" perturbations. Roughly speaking, they consider a continuous-time Itô system of the form

$$dx(t) = Ax(t)dt + \sum_{i=1}^{L} B_i \Delta_i(Cx(t)) dp_i(t),$$

where the operators Δ_i are Lipschitz continuous, with $\Delta_i(0) = 0$; the processes p_i, $i = 1, \ldots, L$, are standard Brownian motions (which, roughly speaking, are integrals of white noise).

We note that our framework can be recovered by assuming that the operators Δ_i are restricted to a diagonal structure of the form $\Delta_i(z) = \sigma_i z$.

Measuring the "size" of the perturbation term by $\left(\sum_{i=1}^{L} \|\Delta_i\|^2\right)^{1/2}$, where $\|\cdot\|$ denotes the Lipschitz norm, leads the authors of [EP92] to the robustness margin

$$\begin{array}{ll} \text{minimize} & \gamma \\ \text{subject to} & P > 0, \quad A^T P + PA + C^T C < 0, \quad B_i^T P B_i \leq \gamma, \quad i = 1, \ldots, L \end{array}$$

For $L = 1$, and B_1 and C are vectors, that is, when a single scalar uncertain parameter perturbs an LTI system, both this result and ours coincide with the $\mathbf{H}_2$ norm of (A, B_1, C). This is consistent with the earlier results of Kleinmann [KLE69] and Willems [WB71] (see also [BB91, P114] for a related result). For $L > 1$ however, applying their results to our framework would be conservative, since we only consider "diagonal" perturbations with "repeated elements", $\Delta_i = \sigma_i I$. Note that the LMI approach can also solve the state-feedback synthesis problem in the framework of [EP92]. El Bouhtouri and Pritchard provide a solution of this problem in [EP94].

Stability conditions for arbitrary noise intensities are given in geometric terms in [WW83]. In our framework, this happens if and only if the optimal value of the corresponding GEVP is zero.

Bounds on state covariance with noise input

We consider system (9.6) in which w is a unit white noise process, i.e., $w(k)$ are independent, identically distributed with $\mathbf{E}\, w(k) = 0$, $\mathbf{E}\, w(k)w(k)^T = W$, and w is independent of p. In this case, of course, the state mean $\bar{x}$ is zero. We derive an upper bound on the state correlation (or covariance) of the system.

The state correlation $M(k)$ satisfies the difference equation

$$M(k+1) = AM(k)A^T + B_w W B_w^T + \sum_{i=1}^{L} \sigma_i^2 \left(A_i M(k) A_i^T + B_{w,i} W B_{w,i}^T \right),$$

with $M(0) = 0$.

Since the system is mean-square stable, the state correlation $M(k)$ has a (finite) limit M_∞ which satisfies

$$M_\infty = AM_\infty A^T + B_w W B_w^T + \sum_{i=1}^{L} \sigma_i^2 \left(A_i M_\infty A_i^T + B_{w,i} W B_{w,i}^T \right).$$

In fact, the matrix M_∞ can be computed directly as the solution of a linear matrix *equation*. However, the above LMI formulation extends immediately to more complicated situations (for instance, when the white noise input w has a covariance matrix with unknown off-diagonal elements), while the "direct" method does not. See the Notes and References for Chapter 6 for details on how these more complicated situations can be handled.

Extension to other stochastic systems

The Lur'e stability problem considered in §8.1 can be extended to the stochastic framework; see Wonham [WON66]. This extension can be cast in terms of LMIs.

It is also possible to extend the results in this chapter to "uncertain stochastic systems", of the form

$$x(k+1) = \left(A(k) + \sum_{i=1}^{L} \sigma_i A_i p_i(k) \right) x(k),$$

where the time-varying matrix $A(k)$ is only known to belong to a given polytope.

State-feedback synthesis

Conditions for stabilizability for arbitrary noise intensities are given in geometric terms in [WW83]. Apart from this work, the bulk of earlier research on this topic concentrated on Linear-Quadratic Regulator theory for continuous-time Itô systems, which addresses the problem of finding an input u that minimize a performance index of the form

$$J(K) = \mathbf{E} \int_0^\infty \left(x(t)^T Q x(t) + u(t)^T R u(t) \right) \, dt$$

where $Q \geq 0$ and $R > 0$ are given. The solution of this problem, as found in [McL71] or [Ben92], can be expressed as a (linear, constant) state-feedback $u(t) = Kx(t)$ where the gain K is given in terms of the solution of a non-standard Riccati equation. (See [Kle69, Won68, FR75, BH88b] for additional details.) The existing methods (see e.g. [Phi89, RSH90]) use homotopy algorithms to solve these equations, with no guarantee of global convergence. We can solve these non-standard Riccati equations reliably (i.e., with guaranteed convergence), by recognizing that the solution is an extremal point of a certain LMI feasibility set.

Chapter 10

Miscellaneous Problems

10.1 Optimization over an Affine Family of Linear Systems

We consider a family of linear systems,

$$\dot{x} = Ax + B_w w, \qquad z = C_z(\theta)x + D_{zw}(\theta)w, \tag{10.1}$$

where C_z and D_{zw} depend affinely on a parameter $\theta \in \mathbf{R}^p$. We assume that A is stable and $(A,\, B_w)$ is controllable. The transfer function,

$$H_\theta(s) \stackrel{\Delta}{=} C_z(\theta)(sI - A)^{-1}B_w + D_{zw}(\theta),$$

depends affinely on θ.

Several problems arising in system and control theory have the form

$$\begin{array}{ll} \text{minimize} & \varphi_0(H_\theta) \\ \text{subject to} & \varphi_i(H_\theta) < \alpha_i, \quad i = 1, \ldots, p \end{array} \tag{10.2}$$

where φ_i are various convex functionals. These problems can often be recast as LMI problems. To do this, we will represent $\varphi_i(H_\theta) < \alpha_i$ as an LMI in θ, α_i, and possibly some auxiliary variables ζ_i (for each i):

$$F_i(\theta, \alpha_i, \zeta_i) > 0.$$

The general optimization problem (10.2) can then be expressed as the EVP

$$\begin{array}{ll} \text{minimize} & \alpha_0 \\ \text{subject to} & F_i(\theta, \alpha_i, \zeta_i) > 0, \quad i = 0, 1, \ldots, p \end{array}$$

10.1.1 $\mathbf{H}_2$ norm

The $\mathbf{H}_2$ norm of the system (10.1), i.e.,

$$\|H_\theta\|_2^2 = \frac{1}{2\pi}\,\mathbf{Tr} \int_0^\infty H_\theta(j\omega)^* H_\theta(j\omega)\, d\omega,$$

is finite if and only if $D_{zw}(\theta) = 0$. In this case, it equals $\mathbf{Tr}\, C_z W_c C_z^T$, where $W_c > 0$ is the controllability Gramian of the system (10.1) which satisfies (6.6). Therefore, the $\mathbf{H}_2$ norm of the system (10.1) is less than or equal to γ if and only if the following conditions on θ and γ^2 are satisfied:

$$D_{zw}(\theta) = 0, \qquad \mathbf{Tr}\, C_z(\theta) W_c C_z(\theta)^T \le \gamma^2.$$

The quadratic constraint on θ is readily cast as an LMI.

10.1.2 $\mathbf{H}_\infty$ norm

From the bounded-real lemma, we have $\|H_\theta\|_\infty < \gamma$ if and only if there exists $P \geq 0$ such that

$$\begin{bmatrix} A^TP + PA & PB_w \\ B_w^TP & -\gamma^2 I \end{bmatrix} + \begin{bmatrix} C_z(\theta)^T \\ D_{zw}(\theta)^T \end{bmatrix} \begin{bmatrix} C_z(\theta) & D_{zw}(\theta) \end{bmatrix} \leq 0.$$

Remark: Here we have converted the so-called "semi-infinite" convex constraint $\|H_\theta(i\omega)\| < \gamma$ for all $\omega \in \mathbf{R}$ into a finite-dimensional convex (linear matrix) inequality.

10.1.3 Entropy

The γ-*entropy* of the system (10.1) is defined as

$$I_\gamma(H_\theta) \stackrel{\Delta}{=} \begin{cases} \dfrac{-\gamma^2}{2\pi} \displaystyle\int_{-\infty}^{\infty} \log\det(I - \gamma^2 H_\theta(i\omega)H_\theta(i\omega)^*)\, d\omega, & \text{if } \|H_\theta\|_\infty < \gamma, \\ \infty, & \text{otherwise.} \end{cases}$$

When it is finite, $I_\gamma(H_\theta)$ is given by $\mathbf{Tr}\, B_w^TPB_w$, where P is a symmetric matrix with the smallest possible maximum singular value among all solutions of the Riccati equation

$$A^TP + PA + C_z(\theta)^TC_z(\theta) + \frac{1}{\gamma^2}PB_wB_w^TP = 0.$$

For the system (10.1), the γ-entropy constraint $I_\gamma \leq \lambda$ is therefore equivalent to an LMI in θ, $P = P^T$, γ^2, and λ:

$$D_{zw}(\theta) = 0, \qquad \begin{bmatrix} A^TP + PA & PB_w & C_z(\theta)^T \\ B_w^TP & -\gamma^2 I & 0 \\ C_z(\theta) & 0 & -I \end{bmatrix} \leq 0, \qquad \mathbf{Tr}\, B_w^TPB_w \leq \lambda.$$

10.1.4 Dissipativity

Suppose that w and z are the same size. The dissipativity of H_θ (see (6.59)) exceeds γ if and only if the LMI in the variables $P = P^T$ and θ holds:

$$\begin{bmatrix} A^TP + PA & PB_w - C_z(\theta)^T \\ B_w^TP - C_z(\theta) & 2\gamma I - D_{zw}(\theta) - D_{zw}(\theta)^T \end{bmatrix} \leq 0.$$

We remind the reader that passivity corresponds to zero dissipativity.

10.1.5 Hankel norm

The Hankel norm of H_θ is less than γ if and only if the following LMI in Q, θ and γ^2 holds:

$$\begin{array}{ll} D_{zw}(\theta) = 0, & A^TQ + QA + C_z(\theta)^TC_z(\theta) \leq 0, \\ \gamma^2 I - W_c^{1/2}QW_c^{1/2} \geq 0, & Q \geq 0. \end{array}$$

(W_c is the controllability Gramian defined by (6.6).)

10.2 Analysis of Systems with LTI Perturbations

In this section, we address the problem of determining the stability of the system

$$\dot{x} = Ax + B_p p, \qquad q = C_q x + D_{qp} p, \qquad p_i = \delta_i * q_i, \quad i = 1, \ldots, n_p, \tag{10.3}$$

where $p = [p_1 \ \cdots \ p_{n_p}]^T$, $q = [q_1 \ \cdots \ q_{n_p}]^T$, and δ_i, $i = 1, \ldots, n_p$, are impulse responses of single-output single-output LTI systems whose $\mathbf{H}_\infty$ norm is strictly less than one, and $*$ denotes convolution. This problem has many variations: Some of the δ_i may be impulse response matrices, or they may satisfy equality constraints such as $\delta_1 = \delta_2$. Our analysis can be readily extended to these cases.

Let $H_{qp}(s) = C_q(sI - A)^{-1}B_p + D_{qp}$. Suppose that there exists a diagonal transfer matrix $W(s) = \mathbf{diag}(W_1(s), \ldots, W_{n_p}(s))$, where $W_1(s), \ldots, W_{n_p}(s)$ are single-input single-output transfer functions with no poles or zeros on the imaginary axis such that

$$\sup_{\omega \in \mathbf{R}} \left\| W(i\omega) H_{qp}(i\omega) W(i\omega)^{-1} \right\| < 1 \tag{10.4}$$

Then from a small-gain argument, the system (10.3) is stable. Such a W is referred to as a *frequency-dependent scaling*.

Condition (10.4) is equivalent to the existence of $\gamma > 0$ such that

$$H_{qp}(i\omega)^* W(i\omega)^* W(i\omega) H_{qp}(i\omega) - W(i\omega)^* W(i\omega) + \gamma I \leq 0,$$

for all $\omega \in \mathbf{R}$. From spectral factorization theory, it can be shown that such a W exists if and only if there exists a stable, diagonal transfer matrix V and positive μ such that

$$\begin{array}{l} H_{qp}(i\omega)^*(V(i\omega) + V(i\omega)^*)H_{qp}(i\omega) - (V(i\omega) + V(i\omega)^*) + \gamma I \leq 0, \\ V(i\omega) + V(i\omega)^* \geq \mu I, \end{array} \tag{10.5}$$

for all $\omega \in \mathbf{R}$.

In order to reduce (10.5) to an LMI condition, we restrict V to belong to an affine, finite-dimensional set Θ of diagonal transfer matrices, i.e., of the form

$$V_\theta(s) = D_{\mathrm{V}}(\theta) + C_{\mathrm{V}}(\theta)(sI - A)^{-1}B,$$

where D_{V} and C_{V} are affine functions are θ.

Then, $H_{qp}(-s)^T V_\theta(s) H_{qp}(s) - V_\theta(s)$ has a state-space realization (A_{aug}, B_{aug}, $C_{\mathrm{aug}}(\theta)$, $D_{\mathrm{aug}}(\theta)$) where C_{aug} and D_{aug} are affine in θ, and ($A_{\mathrm{aug}}, B_{\mathrm{aug}}$) is controllable.

From §10.1.4, the condition that $V_\theta(i\omega) + V_\theta(i\omega)^* \geq \mu I$ for all $\omega \in \mathbf{R}$ is equivalent to the existence of $P_1 \geq 0$, θ and $\gamma > 0$ such that the LMI

$$\begin{bmatrix} A_{\mathrm{V}}^T P_1 + P_1 A_{\mathrm{V}} & P_1 B_{\mathrm{V}} - C_{\mathrm{V}}(\theta)^T \\ (P_1 B_{\mathrm{V}} - C_{\mathrm{V}}(\theta)^T)^T & -(D_{\mathrm{V}}(\theta) + D_{\mathrm{V}}(\theta)^T) + \mu I \end{bmatrix} \leq 0$$

holds.

The condition $H_{qp}(i\omega)^*(V_\theta(i\omega) + V_\theta(i\omega)^*)H_{qp}(i\omega) - (V_\theta(i\omega) + V_\theta(i\omega)^*) + \gamma I \leq 0$ is equivalent to the existence of $P_2 = P_2^T$ and θ satisfying the LMI

$$\begin{bmatrix} A_{\mathrm{aug}}^T P_2 + P_2 A_{\mathrm{aug}} & P_2 B_{\mathrm{aug}} + C_{\mathrm{aug}}(\theta)^T \\ (P_2 B_{\mathrm{aug}} + C_{\mathrm{aug}}(\theta)^T)^T & D_{\mathrm{aug}}(\theta) + D_{\mathrm{aug}}(\theta)^T + \gamma I \end{bmatrix} \leq 0. \tag{10.6}$$

(See the Notes and References for details.) Thus proving stability of the system (10.3) using a finite-dimensional set of frequency-dependent scalings is an LMIP.

10.3 Positive Orthant Stabilizability

The LTI system $\dot{x} = Ax$ is said to be *positive orthant stable* if $x(0) \in \mathbf{R}_+^n$ implies that $x(t) \in \mathbf{R}_+^n$ for all $t \geq 0$, and $x(t) \to 0$ as $t \to \infty$. It can be shown that the system $\dot{x} = Ax$ is positive orthant stable if and only if $A_{ij} \geq 0$ for $i \neq j$, and there exists a diagonal $P > 0$ such that $PA^T + AP < 0$. Therefore checking positive orthant stability is an LMIP.

We next consider the problem of finding K such that the system $\dot{x} = (A + BK)x$ is positive orthant stable. Equivalently, we seek a diagonal $Q > 0$ and K such that $Q(A + BK)^T + (A + BK)Q < 0$, with the off-diagonal entries of $A + BK$ being nonnegative. Since $Q > 0$ is diagonal, the last condition holds if and only if all the off-diagonal entries of $(A + BK)Q$ are nonnegative. Therefore, with the change of variables $Y = KQ$, proving positive orthant stabilizability for the system $\dot{x} = Ax + Bu$ is equivalent to finding a solution to the LMIP with variables Q and Y:

$$Q > 0, \quad (AQ + BY)_{ij} \geq 0, \quad i \neq j, \quad QA^T + Y^T B^T + AQ + BY < 0. \tag{10.7}$$

Remark: The method can be extended to invariance of more general (polyhedral) sets. Positive orthant stability and stabilizability of LDIs can be handled in a similar way. See the Notes and References.

10.4 Linear Systems with Delays

Consider the system described by the delay-differential equation

$$\frac{d}{dt}x(t) = Ax(t) + \sum_{i=1}^{L} A_i x(t - \tau_i), \tag{10.8}$$

where $x(t) \in \mathbf{R}^n$, and $\tau_i > 0$. If the functional

$$V(x,t) = x(t)^T P x(t) + \sum_{i=1}^{L} \int_0^{\tau_i} x(t-s)^T P_i x(t-s)\, ds, \tag{10.9}$$

where $P > 0$, $P_1 > 0, \ldots, P_L > 0$, satisfies $dV(x,t)/dt < 0$ for every x satisfying (10.8), then the system (10.8) is stable, i.e., $x(t) \to 0$ as $t \to \infty$.

It can be verified that $dV(x,t)/dt = y(t)^T W y(t)$, where

$$W = \begin{bmatrix} A^T P + PA + \sum_{i=1}^{L} P_i & PA_1 & \cdots & PA_L \\ A_1^T P & -P_1 & \cdots & 0 \\ \vdots & \vdots & \ddots & \vdots \\ A_L^T P & 0 & \cdots & -P_L \end{bmatrix}, \quad y(t) = \begin{bmatrix} x(t) \\ x(t-\tau_1) \\ \vdots \\ x(t-\tau_L) \end{bmatrix}.$$

Therefore, we can prove stability of system (10.8) using Lyapunov functionals of the form (10.9) by solving the LMIP $W < 0$, $P > 0$, $P_1 > 0, \ldots, P_L > 0$.

Remark: The techniques for proving stability of norm-bound LDIs, discussed in Chapter 5, can also be used on the system (10.8); it can be verified that the LMI (5.15) that we get for quadratic stability of a DNLDI, in the case when the state x is a scalar, is the same as the LMI $W < 0$ above. We also note that we can regard the delay elements of system (10.8) as convolution operators with unit $\mathbf{L}_2$ gain, and use the techniques of §10.2 to check its stability.

Next, we add an input u to the system (10.8) and consider

$$\frac{d}{dt}x(t) = Ax(t) + Bu(t) + \sum_{i=1}^{L} A_i x(t-\tau_i). \tag{10.10}$$

We seek a state-feedback $u(t) = Kx(t)$ such that the system (10.10) is stable.

From the discussion above, there exists a state-feedback gain K such that a Lyapunov functional of the form (10.9) proves the stability of the system (10.10), if

$$W = \begin{bmatrix} (A+BK)^T P + P(A+BK) + \sum_{i=1}^{L} P_i & PA_1 & \cdots & PA_L \\ A_1^T P & -P_1 & \cdots & 0 \\ \vdots & \vdots & \ddots & \vdots \\ A_L^T P & 0 & \cdots & -P_L \end{bmatrix} < 0$$

for some $P, P_1, \ldots, P_L > 0$.

Multiplying every block entry of W on the left and on the right by P^{-1} and setting $Q = P^{-1}$, $Q_i = P^{-1} P_i P^{-1}$ and $Y = KP^{-1}$, we obtain the condition

$$X = \begin{bmatrix} AQ + QA^T + BY + Y^T B^T + \sum_{i=1}^{L} Q_i & A_1 Q & \cdots & A_L Q \\ QA_1^T & -Q_1 & \cdots & 0 \\ \vdots & \vdots & \ddots & \vdots \\ QA_L^T & 0 & \cdots & -Q_L \end{bmatrix} < 0.$$

Thus, checking stabilizability of the system (10.8) using Lyapunov functionals of the form (10.9) is an LMIP in the variables $Q, Y, Q_1, \ldots, Q_L$.

Remark: It is possible to use the elimination procedure of §2.6.2 to eliminate the matrix variable Y and obtain an equivalent LMIP with fewer variables.

10.5 Interpolation Problems

10.5.1 Tangential Nevanlinna-Pick problem

Given $\lambda_1, \ldots, \lambda_m$ with $\lambda_i \in \mathbf{C}_+ \stackrel{\Delta}{=} \{s \mid \mathbf{Re}\, s > 0\}$ and distinct, $u_1, \ldots, u_m$, with $u_i \in \mathbf{C}^q$ and $v_1, \ldots, v_m$, with $v_i \in \mathbf{C}^p$, $i = 1, \ldots, m$, the tangential Nevanlinna-Pick problem is to find, if possible, a function $H : \mathbf{C} \longrightarrow \mathbf{C}^{p\times q}$ which is analytic in $\mathbf{C}_+$, and satisfies

$$H(\lambda_i)u_i = v_i, \quad i = 1, \ldots, m, \text{ with } \|H\|_\infty \le 1. \tag{10.11}$$

Problem (10.11) arises in multi-input multi-output $\mathbf{H}_\infty$ control theory.

From Nevanlinna-Pick theory, problem (10.11) has a solution and only if the Pick matrix N defined by

$$N_{ij} = \frac{u_i^* u_j - v_i^* v_j}{\lambda_i^* + \lambda_j)}$$

is positive semidefinite. N can also be obtained as the solution of the Lyapunov equation

$$A^*N + NA - (U^*U - V^*V) = 0.$$

where $A = \mathbf{diag}(\lambda_1, \ldots, \lambda_m)$, $U = [u_1 \ \cdots \ u_m]$, $V = [v_1 \ \cdots \ v_m]$. For future reference, we note that $N = G_{\rm in} - G_{\rm out}$, where

$$A^*G_{\rm in} + G_{\rm in}A - U^*U = 0, \qquad A^*G_{\rm out} + G_{\rm out}A - V^*V = 0.$$

Solving the tangential Nevanlinna-Pick problem simply requires checking whether $N \ge 0$.

10.5.2 Nevanlinna-Pick interpolation with scaling

We now consider a simple variation of problem (10.11): Given $\lambda_1, \ldots, \lambda_m$ with $\lambda_i \in \mathbf{C}_+$, $u_1, \ldots, u_m$, with $u_i \in \mathbf{C}^p$ and $v_1, \ldots, v_m$, with $v_i \in \mathbf{C}^p$, $i = 1, \ldots, m$, the problem is to find

$$\gamma_{\rm opt} = \inf \left\{ \|DHD^{-1}\|_\infty \;\middle|\; \begin{array}{ll} H \text{ is analytic in } \mathbf{C}_+, & D = D^* > 0 \\ D \in \mathcal{D}, \quad H(\lambda_i)u_i = v_i, & i = 1, \ldots, m \end{array} \right\}, \tag{10.12}$$

where $\mathcal{D}$ is the set of $m \times m$ block-diagonal matrices with some specified block structure. Problem (10.12) corresponds to finding the smallest scaled $\mathbf{H}_\infty$ norm of all interpolants. This problem arises in multi-input multi-output $\mathbf{H}_\infty$ control synthesis for systems with structured perturbations.

With a change of variables $P = D^*D$ and the discussion in the previous section, it follows that $\gamma_{\rm opt}$ is the smallest positive γ such that there exists $P > 0$, $P \in \mathcal{D}$ such that the following equations and inequality hold:

$$\begin{aligned} &A^*G_{\rm in} + G_{\rm in}A - U^*PU = 0, \\ &A^*G_{\rm out} + G_{\rm out}A - V^*PV = 0, \\ &\gamma^2 G_{\rm in} - G_{\rm out} \ge 0. \end{aligned}$$

This is a GEVP.

10.5.3 Frequency response identification

We consider the problem of identifying the transfer function H of a linear system from noisy measurements of its frequency response at a set of frequencies. We seek H satisfying two constraints:

- *Consistency with measurements.* For some n_i with $|n_i| \leq \epsilon$, we have $f_i = H(j\omega_i)+n_i$. Here, f_i is the measurement of the frequency response at frequency ω_i, n_i is the (unknown) measurement error, and ϵ is the measurement precision.
- *Prior assumption.* α-shifted $\mathbf{H}_\infty$ norm of H does not exceed M.

From Nevanlinna-Pick theory, there exist H and n with $H(\omega_i) = f_i + n_i$ satisfying these conditions if and only if there exist n_i with $|n_i| \leq \epsilon$, $G_{\text{in}} > 0$ and $G_{\text{out}} > 0$ such that

$$\begin{aligned} &M^2 G_{\text{in}} - G_{\text{out}} \geq 0, \\ &(A+\alpha I)^* \, G_{\text{in}} + G_{\text{in}} \, (A+\alpha I) - e^* e = 0, \\ &(A+\alpha I)^* \, G_{\text{out}} + G_{\text{out}} \, (A+\alpha I) - (f+n)^*(f+n) = 0, \end{aligned}$$

where $A = \mathbf{diag}(j\omega_1, \ldots, j\omega_m)$, $e = [1 \ \cdots \ 1]$, $f = [f_1 \ \cdots \ f_m]$, and $n = [n_1 \ \cdots \ n_m]$. It can be shown that these conditions are equivalent to

$$\begin{aligned} &M^2 G_{\text{in}} - G_{\text{out}} \geq 0, \\ &(A+\alpha I)^* \, G_{\text{in}} + G_{\text{in}} \, (A+\alpha I) - e^* e = 0, \\ &(A+\alpha I)^* \, G_{\text{out}} + G_{\text{out}} \, (A+\alpha I) - (f+n)^*(f+n) \geq 0, \end{aligned}$$

with $|n_i| \leq \epsilon$.

With this observation, we can answer a number of interesting questions in frequency response identification by solving EVPs and GEVPs.

- *For fixed α and ϵ, minimize M.* Solving this GEVP answers the question "Given α and a bound on the noise values, what is the smallest possible α-shifted $\mathbf{H}_\infty$ norm of the system consistent with the measurements of the frequency response?"
- *For fixed α and M, minimize ϵ.* Solving this EVP answers the question "Given α and a bound on α-shifted $\mathbf{H}_\infty$ norm of the system, what is the "smallest" possible noise such that the measurements are consistent with the given values of α and M.

10.6 The Inverse Problem of Optimal Control

Given a system

$$\dot{x} = Ax + Bu, \quad x(0) = x_0, \qquad z = \begin{bmatrix} Q^{1/2} & 0 \\ 0 & R^{1/2} \end{bmatrix} \begin{bmatrix} x \\ u \end{bmatrix},$$

with (A,B) is stabilizable, (Q,A) is detectable and $R > 0$, the LQR problem is to determine u that minimizes

$$\int_0^\infty z^T z \, dt.$$

The solution of this problem can be expressed as a state-feedback $u = Kx$ with $K = -R^{-1}B^T P$, where P is the unique nonnegative solution of

$$A^T P + PA - PBR^{-1}B^T P + Q = 0.$$

The inverse problem of optimal control is the following. Given a matrix K, determine if there exist $Q \geq 0$ and $R > 0$, such that (Q,A) is detectable and $u = Kx$ is the optimal control for the corresponding LQR problem. Equivalently, we seek $R > 0$ and $Q \geq 0$ such that there exists P nonnegative and P_1 positive-definite satisfying

$$(A+BK)^T P + P(A+BK) + K^T RK + Q = 0, \qquad B^T P + RK = 0$$

and $A^T P_1 + P_1 A < Q$. This is an LMIP in P, P_1, R and Q. (The condition involving P_1 is equivalent to (Q, A) being detectable.)

10.7 System Realization Problems

Consider the discrete-time minimal LTI system

$$x(k+1) = Ax(k) + Bu(k), \qquad y(k) = Cx(k), \tag{10.13}$$

where $x : \mathbf{Z}_+ \to \mathbf{R}^n$, $u : \mathbf{Z}_+ \to \mathbf{R}^{n_u}$ $y : \mathbf{Z}_+ \to \mathbf{R}^{n_y}$. The *system realization problem* is to find a change of state coordinates $x = T\bar{x}$, such that the new realization

$$\bar{x}(k+1) = T^{-1}AT\bar{x}(k) + T^{-1}Bu(k) \qquad y(k) = CT\bar{x}(k), \tag{10.14}$$

satisfies two competing constraints: First, the input-to-state transfer matrix, $T^{-1}(zI-A)^{-1}B$ should be "small", in order to avoid state overflow in numerical implementation; and second, the state-to-output transfer matrix, $C(zI-A)^{-1}T$ should be "small", in order to minimize effects of state quantization at the output.

Suppose we assume that the RMS value of the input is bounded by α and require the RMS value of the state to be less than one. This yields the bound

$$\left\|T^{-1}(zI-A)^{-1}B\right\|_\infty \leq 1/\alpha, \tag{10.15}$$

where the $\mathbf{H}_\infty$ norm of the transfer matrix $H(z)$ is defined as

$$\|H\|_\infty = \sup\left\{\|H(z)\| \mid |z| > 1\right\}.$$

Next, suppose that the state quantization noise is modeled as a white noise sequence $w(k)$ with $\mathbf{E}\, w(k)^T w(k) = \eta$, injected directly into the state, and its effect on the output is measured by its RMS value, which is just η times the $\mathbf{H}_2$ norm of the state-to-output transfer matrix:

$$p_{\rm noise} = \eta \left\|C(zI-A)^{-1}T\right\|_2. \tag{10.16}$$

Our problem is then to compute T to minimize the noise power (10.16) subject to the overflow avoidance constraint (10.15).

The constraint (10.15) is equivalent to the existence of $P > 0$ such that

$$\begin{bmatrix} A^T PA - P + T^{-T}T^{-1} & A^T PB \\ B^T PA & B^T PB - I/\alpha^2 \end{bmatrix} \leq 0.$$

The output noise power can be expressed as

$$p_{\rm noise}^2 = \eta^2 \,\mathbf{Tr}\, T^T W_{\rm o} T,$$

where $W_{\rm o}$ is the observability Gramian of the original system (A, B, C), given by

$$W_{\rm o} = \sum_{k=0}^{\infty} (A^T)^k C^T C A^k.$$

W_o is the solution of the Lyapunov equation

$$A^T W_\mathrm{o} A - W_\mathrm{o} + C^T C = 0.$$

With $X = T^{-T}T^{-1}$, the realization problem is: Minimize $\eta^2 \,\mathbf{Tr}\, W_\mathrm{o} X^{-1}$ subject to $X > 0$ and the LMI (in $P > 0$ and X)

$$\begin{bmatrix} A^T P A - P + X & A^T P B \\ B^T P A & B^T P B - I/\alpha^2 \end{bmatrix} \leq 0. \tag{10.17}$$

This is a convex problem in X and P, and can be transformed into the EVP

$$\begin{array}{ll} \text{minimize} & \eta^2 \,\mathbf{Tr}\, Y W_\mathrm{c} \\ \text{subject to} & (10.17), \quad P > 0, \quad \begin{bmatrix} Y & I \\ I & X \end{bmatrix} \geq 0 \end{array}$$

Similar methods can be used to handle several variations and extensions.

Input u has RMS value bounded componentwise

Suppose the RMS value of each component u_i is less than α (instead of the RMS value of u being less than α) and that the RMS value of the state is still required to be less than one. Then, using the methods of §10.9, an equivalent condition is the LMI (with variables X, P and R):

$$\begin{bmatrix} A^T P A - P + X & A^T P B \\ B^T P A & B^T P B - R/\alpha^2 \end{bmatrix} \leq 0, \qquad P > 0, \tag{10.18}$$

where $R > 0$ is diagonal, with unit trace, so that the EVP (over X, Y, P and R) is

$$\begin{array}{ll} \text{minimize} & \eta^2 \,\mathbf{Tr}\, Y W_\mathrm{c} \\ \text{subject to} & (10.18), \quad P > 0, \quad X > 0 \\ & R > 0, \quad R \text{ is diagonal}, \quad \mathbf{Tr}\, R = 1 \\ & \begin{bmatrix} Y & I \\ I & X \end{bmatrix} \geq 0 \end{array}$$

Minimizing state-to-output $\mathbf{H}_\infty$ norm

Alternatively, we can measure the effect of the quantization noise on the output with the $\mathbf{H}_\infty$ norm, that is, we can choose T to minimize $\|C(zI - A)^{-1}T\|_\infty$ subject to the constraint (10.15). The constraint $\|C(zI - A)^{-1}T\|_\infty \leq \gamma$ is equivalent to the LMI (over $Y > 0$ and $X = T^{-T}T^{-1} > 0$):

$$\begin{bmatrix} A^T Y A - Y + C^T C & A^T Y \\ Y A & Y - X/\alpha^2 \end{bmatrix} \leq 0. \tag{10.19}$$

Therefore choosing T to minimize $\|C(zI - A)^{-1}T\|_\infty$ subject to (10.15) is the GEVP

$$\begin{array}{ll} \text{minimize} & \gamma \\ \text{subject to} & P > 0, \quad Y > 0, \quad X > 0, \quad (10.17), \quad (10.19) \end{array}$$

An optimal state coordinate transformation T is any matrix that satisfies $X_\mathrm{opt} = (TT^T)^{-1}$, where X_opt is an optimal value of X in the GEVP.

10.8 Multi-Criterion LQG

Consider the LTI system given by

$$\dot{x} = Ax + Bu + w, \qquad y = Cx + v, \qquad z = \begin{bmatrix} Q^{1/2} & 0 \\ 0 & R^{1/2} \end{bmatrix} \begin{bmatrix} x \\ u \end{bmatrix},$$

where u is the control input; y is the measured output; z is the exogenous output; w and v are independent white noise signals with constant, positive-definite spectral density matrices W and V respectively; we further assume that $Q \geq 0$ and $R > 0$, that (A, B) is controllable, and that (C, A) and (Q, A) are observable.

The standard Linear-Quadratic Gaussian (LQG) problem is to minimize

$$J_{\rm lqg} = \lim_{t\to\infty} \mathbf{E}\, z(t)^T z(t)$$

over u, subject to the condition that $u(t)$ is measurable on $y(\tau)$ for $\tau \leq t$. The optimal cost is given by

$$J^*_{\rm lqg} = \mathbf{Tr}\left(X_{\rm lqg} U + Q Y_{\rm lqg}\right),$$

where $X_{\rm lqg}$ and $Y_{\rm lqg}$ are the unique positive-definite solutions of the Riccati equations

$$\begin{array}{l} A^T X_{\rm lqg} + X_{\rm lqg} A - X_{\rm lqg} B R^{-1} B^T X_{\rm lqg} + Q = 0, \\ A Y_{\rm lqg} + Y_{\rm lqg} A^T - Y_{\rm lqg} C^T V^{-1} C Y_{\rm lqg} + W = 0, \end{array} \tag{10.20}$$

and $U = Y_{\rm lqg} C^T V^{-1} C Y_{\rm lqg}$.

In the multi-criterion LQG problem, we have several exogenous outputs of interest given by

$$z_i = \begin{bmatrix} Q_i^{1/2} & 0 \\ 0 & R_i^{1/2} \end{bmatrix} \begin{bmatrix} x \\ u \end{bmatrix}, \qquad Q_i \geq 0, \qquad R_i > 0, \quad i = 0, \ldots, p.$$

We assume that (Q_0, A) is observable. For each z_i, we associate a cost function

$$J^i_{\rm lqg} = \lim_{t\to\infty} \mathbf{E}\, z_i(t)^T z_i(t), \quad i = 0, \ldots, p. \tag{10.21}$$

The multi-criterion LQG problem is to minimize $J^0_{\rm lqg}$ over u subject to the measurability condition and the constraints $J^i_{\rm lqg} < \gamma_i$, $i = 1, \ldots, p$. This is a convex optimization problem, whose solution is given by maximizing

$$\mathbf{Tr}\left(X_{\rm lqg} U + Q Y_{\rm lqg}\right) - \sum_{i=1}^p \gamma_i \tau_i,$$

over nonnegative $\tau_1, \ldots, \tau_p$, where $X_{\rm lqg}$ and $Y_{\rm lqg}$ are the solutions of (10.20) with

$$Q = Q_0 + \sum_{i=1}^p \tau_i Q_i \qquad \text{and} \qquad R = R_0 + \sum_{i=1}^p \tau_i R_i.$$

Noting that $X_{\rm lqg} \geq X$ for every $X > 0$ that satisfies

$$A^T X + XA - XBR^{-1}B^T X + Q \geq 0,$$

we conclude that the optimal cost is the maximum of

$$\mathbf{Tr}\left(XU + (Q_0 + \sum_{i=1}^p \tau_i Q_i) Y_{\rm lqg}\right) - \sum_{i=1}^p \gamma_i \tau_i,$$

over X, $\tau_1, \ldots, \tau_p$, subject to $X > 0$, $\tau_1 \geq 0, \ldots, \tau_p \geq 0$ and

$$A^T X + XA - XB\left(R_0 + \sum_{i=1}^{p} \tau_i R_i\right)^{-1} B^T X + Q_0 + \sum_{i=1}^{p} \tau_i Q_i \geq 0.$$

Computing this is an EVP.

10.9 Nonconvex Multi-Criterion Quadratic Problems

In this section, we consider the LTI system $\dot{x} = Ax + Bu$, $x(0) = x_0$, where (A, B) is controllable. For any u, we define a set of $p+1$ cost indices $J_0, \ldots, J_p$ by

$$J_i(u) = \int_0^\infty \begin{bmatrix} x^T & u^T \end{bmatrix} \begin{bmatrix} Q_i & C_i \\ C_i^T & R_i \end{bmatrix} \begin{bmatrix} x \\ u \end{bmatrix} dt, \quad i = 0, \ldots, p.$$

Here the symmetric matrices

$$\begin{bmatrix} Q_i & C_i \\ C_i^T & R_i \end{bmatrix}, \quad i = 0, \ldots, p,$$

are *not* necessarily positive-definite. The constrained optimal control problem is:

$$\begin{array}{ll} \text{minimize} & J_0, \\ \text{subject to} & J_i \leq \gamma_i, \quad i = 1, \ldots, p, \qquad x(t) \to 0 \text{ as } t \to \infty \end{array} \tag{10.22}$$

The solution to this problem proceeds as follows: We first define

$$Q = Q_0 + \sum_{i=1}^{p} \tau_i Q_i, \qquad R = R_0 + \sum_{i=1}^{p} \tau_i R_i, \qquad C = C_0 + \sum_{i=1}^{p} \tau_i C_i,$$

where $\tau_1 \geq 0, \ldots, \tau_p \geq 0$. Next, with $\tau = [\tau_1 \ \cdots \ \tau_p]$, we define

$$S(\tau, u) \triangleq J_0 + \sum_{i=1}^{p} \tau_i J_i - \sum_{i=1}^{p} \tau_i \gamma_i.$$

Then, the solution of problem (10.22), when it is not $-\infty$, is given by:

$$\sup_{\tau} \inf_{u} \{ S(\tau, u) \mid x(t) \to 0 \text{ as } t \to \infty \}$$

(See the Notes and References.)

For any fixed τ, the infimum over u of $S(\tau, u)$, when it is not $-\infty$, is computed by solving the EVP in $P = P^T$

$$\begin{array}{ll} \text{maximize} & x(0)^T P x(0) - \sum_{i=1}^{p} \tau_i \gamma_i \\ \text{subject to} & \begin{bmatrix} A^T P + PA + Q & PB + C^T \\ B^T P + C & R \end{bmatrix} \geq 0 \end{array} \tag{10.23}$$

Therefore, the optimal control problem is solved by the EVP in $P = P^T$ and τ:

$$\begin{array}{ll} \text{maximize} & x(0)^T P x(0) - \sum_{i=1}^{p} \tau_i \gamma_i \\ \text{subject to} & \begin{bmatrix} A^T P + PA + Q & PB + C^T \\ B^T P + C & R \end{bmatrix} \geq 0, \qquad \tau_1 \geq 0, \ldots, \tau_p \geq 0. \end{array}$$

Notes and References

Optimization over an affine family of linear systems

Optimization over an affine family of transfer matrices arises in linear controller design via Youla's parametrization of closed-loop transfer matrices; see Boyd and Barratt [BB91] and the references therein. Other examples include finite-impulse-response (FIR) filter design (see for example [OKUI88]) and antenna array weight design [KRB89].

In [KAV94] Kavranoğlu considers the problem of approximating in the $\mathbf{H}_\infty$ norm a given transfer matrix by a constant matrix, which is a problem of the form considered in §10.1. He casts the problem as an EVP.

Stability of systems with LTI perturbations

For references about this problem, we refer the reader to the Notes and References in Chapter 3. LMI (10.6) is derived from Theorems 3 and 4 in Willems [WIL71B].

Positive orthant stabilizability

The problem of positive orthant stability and holdability was extensively studied by Berman, Neumann and Stern [BNS89, §7.4]. They solve the positive orthant holdability problem when the control input is scalar; our results extend theirs to the multi-input case. The study of positive orthant holdability draws heavily from the theory of positive matrices, references for which are [BNS89] and the book by Berman and Plemmons [BP79, CH6].

Diagonal solutions to the Lyapunov equation play a central role in the positive orthant stabilizability and holdability problems. A related paper is by Geromel [GER85], who gives a computational procedure to find diagonal solutions of the Lyapunov equation. Diagonal quadratic Lyapunov functions also arise in the study of large-scale systems [MH78]; see also §3.4. A survey of results and applications of diagonal Lyapunov stability is given in Kaszkurewicz and Bhaya [KB93].

A generalization of the concept of positive orthant stability is that of invariance of polyhedral sets. The problem of checking whether a given polyhedron is invariant as well as the associated state-feedback synthesis problem have been considered by many authors, see e.g., [BIT88, BBT90, HB91B] and references therein. The LMI method of §10.3 can be extended to the polyhedral case, using the approach of [CH93]. Finally, we note that the results of §10.3 are easily generalized to the LDIs considered in Chapter 4.

Stabilization of systems with time delays

The introduction of Lyapunov functionals of the form (10.9) is due to Krasovskii [KRA56]. Skorodinskii [SKO90] observed that the problem of proving stability of a system with delays via Lyapunov–Krasovskii functionals is a convex problem. See also [PF92, WBL92, SL93B, FBB92].

It is interesting to note that the "structured stability problem" or "μ-analysis problem" can be cast in terms of linear systems with delays. Consider a stable LTI system with transfer matrix H and L inputs and outputs, connected in feedback with a diagonal transfer matrix Δ with $\mathbf{H}_\infty$ norm less than one. This system is stable for all such Δ if and only if $\sup_{\omega\in\mathbf{R}} \mu(H(j\omega)) < 1$, where μ denotes the structured singular value (see [DOY82]). It can be shown (see [BOY86]) that if the feedback system is stable for all Δ which are diagonal, with $\Delta_{ii}(s) = e^{-sT_i}$, $T_i > 0$ (i.e., delays) then the system is stable for all Δ with $\|\Delta\|_\infty \le 1$. Thus, verifying $\sup_{\omega\in\mathbf{R}} \mu(H(j\omega)) < 1$ can be cast as checking stability of a linear system with L arbitrary, positive delays. In particular we see that Krasovskii's method from 1956 can be interpreted as a Lyapunov-based μ-analysis method.

Interpolation problems

The classical Nevanlinna-Pick problem dates back at least to 1916 [PIC16, NEV19]. There are two matrix versions of this problem: the matrix Nevanlinna-Pick problem [DGK79] and the tangential Nevanlinna-Pick problem [FED75]. A comprehensive description of the Nevanlinna-Pick problem, its extensions and variations can be found in the book by Ball, Gohberg and Rodman [BGR90].

A number of researchers have studied the application of Nevanlinna-Pick theory to problems in system and control: Delsarte, Genin and Kamp discuss the role of the matrix Nevanlinna-Pick problem in circuit and systems theory [DGK81]. Zames and Francis, in [ZF83], study the implications of interpolation conditions in controller design; also see [OF85]. Chang and Pearson [CP84] solve a class of $\mathbf{H}_\infty$-optimal controller design problems using matrix Nevanlinna-Pick theory. Safonov [SAF86] reduces the design of controllers for systems with structured perturbations to the scaled matrix Nevanlinna-Pick problem. In [KIM87], Kimura reduces a class of $\mathbf{H}_\infty$-optimal controller design problems to the tangential Nevanlinna-Pick problem; see also [BAL94].

System identification

The system identification problem considered in §10.5.3 can be found in [CNF92]. Other system identification problems can be cast in terms of LMI problems; see e.g., [PKT+94].

The inverse problem of optimal control

This problem was first considered by Kalman [KAL64], who solved it when the control u is scalar; see also [AND66C]. It is discussed in great detail by Anderson and Moore [AM90, §5.6], Fujii and Narazaki [FN84] and the references therein; they solve the problem when the control weighting matrix R is known, by checking that the return difference inequality

$$\left(I - B^T(-i\omega I - A^T)^{-1}K\right) R \left(I - K^T(i\omega I - A)^{-1}B\right) \geq R$$

holds for all $\omega \in \mathbf{R}$. Our result handles the case of unknown R, i.e., the most general form of the inverse optimal control problem.

System realization problems

For a discussion of the problem of system realizations see [MR76, THI84], which give analytical solutions via balanced realizations for special realization problems. The book by Gevers and Li [GL93A] describes a number of digital filter realization problems. Liu, Skelton and Grigoriadis discuss the problem of optimal finite word-length controller implementation [LSG92]. See also Rotea and Williamson [RW94B, RW94A].

We conjecture that the LMI approach to system realization can also incorporate the constraint of stability, i.e., that the (nonlinear) system not exhibit limit cycles due to quantization or overflow; see, e.g., [WIL74B].

We also note that an $\mathbf{H}_\infty$ norm constraint on the input-to-state transfer matrix yields, indirectly, a bound on the peak gain from input to state; Boyd and Doyle [BD87]. This can be used to guarantee no overflow given a maximum peak input level.

Multi-criterion LQG

The multi-criterion LQG problem is discussed in [BB91]; see also the references cited there. Rotea [ROT93] calls this problem "generalized $\mathbf{H}_2$ control". He shows that one can restrict attention to observer-based controllers such as

$$\frac{d}{dt}\hat{x} = (A + BK)\hat{x} + YC^TV^{-1}(y - C\hat{x}), \qquad u = K\hat{x},$$

where Y is given by (10.20). The covariance of the residual state $\hat{x}-x$ is Y and the covariance P of the state $\hat{x}$ is given by the solution of the Lyapunov equation:

$$(A+BK)P+P(A+BK)^T+YC^TV^{-1}CY=0. \tag{10.24}$$

The value of each cost index J^i_{lqg} defined by (10.21) is then

$$J^i_{\mathrm{lqg}} = \mathbf{Tr}\, R_i KPK^T + \mathbf{Tr}\, Q_i\,(Y+P).$$

Now, taking $Z = KP$ as a new variable, we replace (10.24) by the linear equality in W and P

$$AP+PA^T+BZ+Z^TB^T+YC^TV^{-1}CY=0$$

and each cost index becomes

$$J^i_{\mathrm{lqg}} = \mathbf{Tr}\, R_i ZP^{-1}Z^T + \mathbf{Tr}\, Q_i\,(Y+P).$$

The multi-criterion problem is then solved by the EVP in P, Z and X

$$\begin{array}{ll} \text{minimize} & \mathbf{Tr}\, R_0 X + \mathbf{Tr}\, Q_0(P+Y) \\ \text{subject to} & \left[\begin{array}{cc} X & Z \\ Z^T & P \end{array}\right] \geq 0, \\ & AP+PA^T+BZ+Z^TB^T+YC^TV^{-1}CY=0, \\ & \mathbf{Tr}\, R_i X + \mathbf{Tr}\, Q_i(P+Y) \leq \gamma_i, \quad i=1,\ldots,L \end{array}$$

This formulation of the multi-criterion LQG problem is the dual of our formulation; the two formulations can be used together in an efficient primal-dual method (see, e.g., [VB93B]). We also mention an algorithm devised by Zhu, Rotea and Skelton [ZRS93] to solve a similar class of problems. For a variation on the multi-criterion LQG problem discussed here, see [TOI84]; see also [TM89, MAK91].

Nonconvex multi-criterion quadratic problems

This section is based on Megretsky [MEG92A, MEG92B, MEG93] and Yakubovich [YAK92]. In our discussion we omitted important technical details, such as constraint regularity, which is covered in these articles. Also, Yakubovich gives conditions under which the infimum is actually a minimum (that is, there exists an optimal control law that achieves the best performance). The EVP (10.23) is derived from Theorem 3 in Willems [WIL71B].

Many of the sufficient conditions for NLDIs in Chapters 5–7 can be shown to be necessary and sufficient conditions when we consider integral constraints on p and q, by analyzing them as nonconvex multi-criterion quadratic problems. This is illustrated in the following section.

Mixed $\mathbf{H}_2$–$\mathbf{H}_\infty$ problem

We consider the analog of an NLDI with integral quadratic constraint:

$$\dot{x} = Ax + B_p p + B_u u, \qquad q = C_q x, \qquad z = \left[\begin{array}{cc} Q^{1/2} & 0 \\ 0 & R^{1/2} \end{array}\right] \left[\begin{array}{c} x \\ u \end{array}\right]$$

$$\int_0^\infty p^T p\, dt \leq \int_0^\infty q^T q\, dt$$

Define the cost function $J = \int_0^\infty z^T z\, dt$. Our goal is to determine $\min_K \max_p J$, over all possible linear state-feedback strategies $u = Kx$. (Note that this is the same as the problem considered in §7.4.1 with integral constraints on p and q instead of pointwise in time constraints.)

For fixed K, finding the maximum of J over admissible p's is done using the results of §10.9. This maximum is obtained by solving the EVP

$$\begin{array}{ll} \text{minimize} & x(0)^T P x(0) \\ \text{subject to} & \begin{bmatrix} \begin{array}{c} (A+B_uK)^T P + P(A+B_uK) + \\ Q + K^T R K + \tau C_q^T C_q \end{array} & PB_p \\ B_p^T P & -\tau I \end{bmatrix} \leq 0, \qquad \tau \geq 0 \end{array}$$

The minimization over K is now simply done by introducing the new variables $W = P^{-1}$, $Y = KW$, $\mu = 1/\tau$. Indeed, in this particular case, we know we can add the constraint $P > 0$. The corresponding EVP is then

$$\begin{array}{ll} \text{minimize} & x(0)^T W^{-1} x(0) \\ \text{subject to} & \begin{bmatrix} \left(\begin{array}{c} WA^T + YB_u^T + AW + B_uY + \\ WQW + Y^T RY + WC_q^T C_q W/\mu \end{array} \right) & \mu B_p \\ \mu B_p^T & -\mu I \end{bmatrix} \leq 0, \qquad \mu \geq 0 \end{array}$$

which is the same as the EVP (7.23).

Notation

$\mathbf{R}$, $\mathbf{R}^k$, $\mathbf{R}^{m\times n}$	The real numbers, real k-vectors, real $m \times n$ matrices.
$\mathbf{R}_+$	The nonnegative real numbers.
$\mathbf{C}$	The complex numbers.
$\mathbf{Re}(a)$	The real part of $a \in \mathbf{C}$, i.e., $(a + a^*)/2$.
$\overline{S}$	The closure of a set $S \subseteq \mathbf{R}^n$.
I_k	The $k \times k$ identity matrix. The subscript is omitted when k is not relevant or can be determined from context.
M^T	Transpose of a matrix M: $(M^T)_{ij} = M_{ji}$.
M^*	Complex-conjugate transpose of a matrix M: $(M^*)_{ij} = M_{ji}^*$, where α^* denotes the complex-conjugate of $\alpha \in \mathbf{C}$.
$\mathbf{Tr}\, M$	Trace of $M \in \mathbf{R}^{n\times n}$, i.e., $\sum_{i=1}^n M_{ii}$.
$M \geq 0$	M is symmetric and positive semidefinite, i.e., $M = M^T$ and $z^T M z \geq 0$ for all $z \in \mathbf{R}^n$.
$M > 0$	M is symmetric and positive-definite, i.e., $z^T M z > 0$ for all nonzero $z \in \mathbf{R}^n$.
$M > N$	M and N are symmetric and $M - N > 0$.
$M^{1/2}$	For $M > 0$, $M^{1/2}$ is the unique $Z = Z^T$ such that $Z > 0$, $Z^2 = M$.
$\lambda_{\max}(M)$	The maximum eigenvalue of the matrix $M = M^T$.
$\lambda_{\max}(P, Q)$	For $P = P^T$, $Q = Q^T > 0$, $\lambda_{\max}(P, Q)$ denotes the maximum eigenvalue of the symmetric pencil (P, Q), i.e., $\lambda_{\max}(Q^{-1/2} P Q^{-1/2})$.
$\lambda_{\min}(M)$	The minimum eigenvalue of $M = M^T$.
$\|M\|$	The spectral norm of a matrix or vector M, i.e., $\sqrt{\lambda_{\max}(M^T M)}$. Reduces to the Euclidean norm, i.e., $\|x\| = \sqrt{x^T x}$, for a vector x.
$\mathbf{diag}(\cdots)$	Block-diagonal matrix formed from the arguments, i.e.,

$$\mathbf{diag}\,(M_1, \ldots, M_m) \triangleq \begin{bmatrix} M_1 & & \\ & \ddots & \\ & & M_m \end{bmatrix}$$

Co S Convex hull of the set $S \subseteq \mathbf{R}^n$, given by

$$\mathbf{Co}\, S \triangleq \left\{ \sum_{i=1}^{p} \lambda_i x_i \,\middle|\, x_i \in S,\ p \geq 0 \right\}.$$

(Without loss of generality, we can take $p = n + 1$ here.)

E x Expected value of (the random variable) x.

Acronyms

Acronym	Meaning	Page
ARE	Algebraic Riccati Equation	3
CP	Convex Minimization Problem	11
DI	Differential Inclusion	51
DNLDI	Diagonal Norm-bound Linear Differential Inclusion	54
EVP	Eigenvalue Minimization Problem	10
GEVP	Generalized Eigenvalue Minimization Problem	10
LDI	Linear Differential Inclusion	51
LMI	Linear Matrix Inequality	7
LMIP	Linear Matrix Inequality Problem	9
LQG	Linear-Quadratic Gaussian	150
LQR	Linear-Quadratic Regulator	114
LTI	Linear Time-invariant	52
NLDI	Norm-bound Linear Differential Inclusion	53
PLDI	Polytopic Linear Differential Inclusion	53
PR	Positive-Real	25
RMS	Root-Mean-Square	91

Bibliography

[ABG93] D. Arzelier, J. Bernussou, and G. Garcia. Pole assignment of linear uncertain systems in a sector via a Lyapunov-type approach. *IEEE Trans. Aut. Control*, AC-38(7):1128–1131, July 1993.

[ABJ75] B. D. Anderson, N. K. Bose, and E. I. Jury. Output feedback stabilization and related problems—Solution via decision methods. *IEEE Trans. Aut. Control*, AC-20:53–66, 1975.

[AC84] J. P. Aubin and A. Cellina. *Differential Inclusions*, volume 264 of *Comprehensive Studies in Mathematics*. Springer-Verlag, 1984.

[AF90] J. P. Aubin and H. Frankowska. *Set-valued analysis*, volume 2 of *Systems & Control*. Birkhauser, Boston, 1990.

[AG64] M. A. Aizerman and F. R. Gantmacher. *Absolute stability of regulator systems*. Information Systems. Holden-Day, San Francisco, 1964.

[AGB94] P. Apkarian, P. Gahinet, and G. Becker. Self-scheduled $\mathbf{H}_\infty$ control of linear parameter-varying systems. Submitted., 1994.

[AGG94] F. Amato, F. Garofalo, and L. Glielmo. Polytopic coverings and robust stability analysis via Lyapunov quadratic forms. In A. Zinober, editor, *Variable structure and Lyapunov control*, volume 193 of *Lecture notes in control and information sciences*, chapter 13, pages 269–288. Springer-Verlag, 1994.

[AKG84] M. Akgül. *Topics in Relaxation and Ellipsoidal Methods*, volume 97 of *Research Notes in Mathematics*. Pitman, 1984.

[AL84] W. F. Arnold and A. J. Laub. Generalized eigenproblem algorithms and software for algebraic Riccati equations. *Proc. IEEE*, 72(12):1746–1754, December 1984.

[ALI91] F. Alizadeh. *Combinatorial optimization with interior point methods and semi-definite matrices*. PhD thesis, Univ. of Minnesota, October 1991.

[ALI92A] F. Alizadeh. Combinatorial optimization with semidefinite matrices. In *Proceedings of second annual Integer Programming and Combinatorial Optimization conference*. Carnegie-Mellon University, 1992.

[ALI92B] F. Alizadeh. Optimization over the positive-definite cone: interior point methods and combinatorial applications. In Panos Pardalos, editor, *Advances in Optimization and Parallel Computing*. North-Holland, 1992.

[ALL89] J. Allwright. On maximizing the minimum eigenvalue of a linear combination of symmetric matrices. *SIAM J. on Matrix Analysis and Applications*, 10:347–382, 1989.

[AM74] B. Anderson and P. J. Moylan. Spectral factorization of a finite-dimensional nonstationary matrix covariance. *IEEE Trans. Aut. Control*, AC-19(6):680–692, 1974.

[AM90] B. Anderson and J. B. Moore. *Optimal Control: Linear Quadratic Methods*. Prentice-Hall, 1990.

[AND66A] B. Anderson. Algebraic description of bounded real matrices. *Electron. Lett.*, 2(12):464–465, December 1966.

[AND66B] B. Anderson. Stability of control systems with multiple nonlinearities. *J. Franklin Inst.*, 282:155–160, 1966.

[AND66C] B. Anderson. The testing for optimality of linear systems. *Int. J. Control*, 4(1):29–40, 1966.

[AND67] B. Anderson. A system theory criterion for positive real matrices. *SIAM J. Control*, 5:171–182, 1967.

[AND73] B. Anderson. Algebraic properties of minimal degree spectral factors. *Automatica*, 9:491–500, 1973. Correction in 11:321–322, 1975.

[ARN74] L. Arnold. *Stochastic differential equations: theory and applications.* Wiley, New York, 1974.

[AS79] M. Araki and M. Saeki. Multivariable circle criteria and M-matrix condition. In *Proc. Joint American Control Conf.*, pages 11–17, 1979.

[ÅST70] K. J. Åström. *Introduction to stochastic control theory*, volume 70 of *Math. in Science and Engr.* Academic Press, 1970.

[AV73] B. Anderson and S. Vongpanitlerd. *Network analysis and synthesis: a modern systems theory approach.* Prentice-Hall, 1973.

[BAD82] F. A. Badawi. On the quadratic matrix inequality and the corresponding algebraic Riccati equation. *Int. J. Control*, 36(2):313–322, August 1982.

[BAL94] V. Balakrishnan. $\mathbf{H}_\infty$ controller synthesis with time-domain constraints using interpolation theory and convex optimization. *IEEE Trans. Aut. Control*, 1994. Submitted.

[BAR70A] E. A. Barbashin. *Introduction to the theory of stability.* Walters-Noordhoff, 1970.

[BAR70B] A. I. Barkin. Sufficient conditions for the absence of auto-oscillations in pulse systems. *Aut. Remote Control*, 31:942–946, 1970.

[BAR83] B. R. Barmish. Stabilization of uncertain systems via linear control. *IEEE Trans. Aut. Control*, AC-28(8):848–850, August 1983.

[BAR85] B. R. Barmish. Necessary and sufficient conditions for quadratic stabilizability of an uncertain system. *Journal of Optimization Theory and Applications*, 46(4):399–408, August 1985.

[BAR93] B. R. Barmish. *New Tools for Robustness of Linear Systems.* MacMillan, 1993.

[BAU63] F. L. Bauer. Optimally scaled matrices. *Numerische Mathematik*, 5:73–87, 1963.

[BB91] S. Boyd and C. Barratt. *Linear Controller Design: Limits of Performance.* Prentice-Hall, 1991.

[BB92] V. Balakrishnan and S. Boyd. Global optimization in control system analysis and design. In C. T. Leondes, editor, *Control and Dynamic Systems: Advances in Theory and Applications*, volume 53. Academic Press, New York, New York, 1992.

[BBB91] V. Balakrishnan, S. Boyd, and S. Balemi. Branch and bound algorithm for computing the minimum stability degree of parameter-dependent linear systems. *Int. J. of Robust and Nonlinear Control*, 1(4):295–317, October–December 1991.

[BBC90] G. Belforte, B. Bona, and V. Cerone. Parameter estimation algorithms for a set-membership description of uncertainty. *Automatica*, 26:887–898, 1990.

[BBFE93] S. Boyd, V. Balakrishnan, E. Feron, and L. El Ghaoui. Control system analysis and synthesis via linear matrix inequalities. In *Proc. American Control Conf.*, volume 2, pages 2147–2154, San Francisco, California, June 1993.

[BBK89] S. Boyd, V. Balakrishnan, and P. Kabamba. A bisection method for computing the $\mathbf{H}_\infty$ norm of a transfer matrix and related problems. *Mathematics of Control, Signals, and Systems*, 2(3):207–219, 1989.

[BBP78] G. P. Barker, A. Berman, and R. J. Plemmons. Positive diagonal solutions to the Lyapunov equations. *Linear and Multilinear Algebra*, 5:249–256, 1978.

[BBT90] C. Burgat, A. Benzaouia, and S. Tarbouriech. Positively invariant sets of discrete-time systems with constrained inputs. *Int. J. Systems Science*, 21:1249–1271, 1990.

[BC85] S. Boyd and L. O. Chua. Fading memory and the problem of approximating nonlinear operators with Volterra series. *IEEE Trans. Circuits Syst.*, CAS-32(11):1150–1171, November 1985.

[BCL83] B. R. Barmish, M. Corless, and G. Leitmann. A new class of stabilizing controllers for uncertain dynamical systems. *SIAM Journal on Control and Optimization*, 21(2):246–255, March 1983.

[BD85] V. G. Borisov and S. N. Diligenskii. Numerical method of stability analysis of nonlinear systems. *Automation and Remote Control*, 46:1373–1380, 1985.

[BD87] S. Boyd and J. Doyle. Comparison of peak and RMS gains for discrete-time systems. *Syst. Control Letters*, 9:1–6, June 1987.

[BDG+91] G. Balas, J. C. Doyle, K. Glover, A. Packard, and Roy Smith. *μ-analysis and synthesis.* MUSYN, inc., and The Mathworks, Inc., 1991.

[BE93] S. Boyd and L. El Ghaoui. Method of centers for minimizing generalized eigenvalues. *Linear Algebra and Applications, special issue on Numerical Linear Algebra Methods in Control, Signals and Systems*, 188:63–111, July 1993.

[BEC91] C. Beck. Computational issues in solving LMIs. In *Proc. IEEE Conf. on Decision and Control*, volume 2, pages 1259–1260, Brighton, December 1991.

[BEC93] G. Becker. *Quadratic Stability and Performance of Linear Parameter Dependent Systems.* PhD thesis, University of California, Berkeley, 1993.

[BEFB93] S. Boyd, L. El Ghaoui, E. Feron, and V. Balakrishnan. Linear matrix inequalities in system and control theory. In *Proc. Annual Allerton Conf. on Communication, Control and Computing*, Allerton House, Monticello, Illinois, October 1993.

[BEN92] A. Bensoussan. *Stochastic Control of Partially Observable Systems.* Cambridge University Press, Cambridge, Great Britain, 1992.

[BF63] R. Bellman and K. Fan. On systems of linear inequalities in Hermitian matrix variables. In V. L. Klee, editor, *Convexity*, volume 7 of *Proceedings of Symposia in Pure Mathematics*, pages 1–11. American Mathematical Society, 1963.

[BFBE92] V. Balakrishnan, E. Feron, S. Boyd, and L. El Ghaoui. Computing bounds for the structured singular value via an interior point algorithm. In *Proc. American Control Conf.*, volume 4, pages 2195–2196, Chicago, June 1992.

[BG94] V. Blondel and M. Gevers. Simultaneous stabilizability question of three linear systems is rationally undecidable. *Mathematics of Control, Signals, and Systems*, 6(1):135–145, 1994.

[BGR90] J. A. Ball, I. Gohberg, and L. Rodman. *Interpolation of Rational Matrix Functions*, volume 45 of *Operator Theory: Advances and Applications.* Birkhäuser, Basel, 1990.

[BGT81] R. G. Bland, D. Goldfarb, and M. J. Todd. The ellipsoid method: A survey. *Operations Research*, 29(6):1039–1091, 1981.

[BH88A] D. Bernstein and W. Haddad. Robust stability and performance analysis for state space systems via quadratic Lyapunov bounds. In *Proc. IEEE Conf. on Decision and Control*, pages 2182–2187, 1988.

[BH88B] D. S. Bernstein and D. C. Hyland. Optimal projection equations for reduced-order modeling, estimation, and control of linear systems with multiplicative noise. *J. Optimiz. Theory Appl.*, 58:387–409, 1988.

[BH89] D. Bernstein and W. Haddad. Robust stability and performance analysis for linear dynamic systems. *IEEE Trans. Aut. Control*, AC-34(7):751–758, July 1989.

[BH93] D. Bernstein and W. Haddad. Nonlinear controllers for positive-real systems with arbitrary input nonlinearities. In *Proc. American Control Conf.*, volume 1, pages 832–836, San Francisco, California, June 1993.

[BHL89] A. C. Bartlett, C. V. Hollot, and H. Lin. Root locations of an entire polytope of polynomials: it suffices to check the edges. *Mathematics of Control, Signals, and Systems*, 1(1):61–71, 1989.

[BHPD94] V. Balakrishnan, Y. Huang, A. Packard, and J. Doyle. Linear matrix inequalities in analysis with multipliers. In *Proc. American Control Conf.*, 1994. Invited session on *Recent Advances in the Control of Uncertain Systems.*

[BIT88] G. Bitsoris. Positively invariant polyhedral sets of discrete-time linear systems. *Int. J. Control*, 47(6):1713–1726, June 1988.

[BJL89] W. W. Barrett, C. R. Johnson, and M. Lundquist. Determinantal formulation for matrix completions associated with chordal graphs. *Linear Algebra and Appl.*, 121:265–289, 1989.

[BK93] B. R. Barmish and H. I. Kang. A survey of extreme point results for robustness of control systems. *Automatica*, 29(1):13–35, 1993.

[BNS89] A. Berman, M. Neumann, and R. Stern. *Nonnegative Matrices in Dynamic Systems.* Wiley-Interscience, 1989.

[BOY86] S. Boyd. A note on parametric and nonparametric uncertainties in control systems. In *Proc. American Control Conf.*, pages 1847–1849, Seattle, Washington, June 1986.

[BOY94] S. Boyd. Robust control tools: Graphical user-interfaces and LMI algorithms. *Systems, Control and Information*, 38(3):111–117, March 1994. Special issue on Numerical Approaches in Control Theory.

[BP79] A. Berman and R. J. Plemmons. *Nonnegative matrices in the mathematical sciences.* Computer Science and Applied Mathematics. Academic Press, 1979.

[BP91] G. Becker and A. Packard. Gain-scheduled state-feedback with quadratic stability for uncertain systems. In *Proc. American Control Conf.*, pages 2702–2703, June 1991.

[BPG89A] J. Bernussou, P. L. D. Peres, and J. C. Geromel. A linear programming oriented procedure for quadratic stabilization of uncertain systems. *Syst. Control Letters*, 13:65–72, 1989.

[BPG89B] J. Bernussou, P. L. D. Peres, and J. C. Geromel. Robust decentralized regulation: A linear programming approach. *IFAC/IFORS/IMACS Symposium Large Scale Systems: Theory and Applications, Berlin-GDR*, 1:135–138, 1989.

[BPPB93] G. Becker, A. Packard, D. Philbrick, and G. Balas. Control of parametrically dependent linear systems: a single quadratic Lyapunov approach. In *1993 American Control Conference*, volume 3, pages 2795–2799, June 1993.

[BR71] D. Bertsekas and I. Rhodes. Recursive state estimation for a set-membership description of uncertainty. *IEEE Trans. Aut. Control*, AC-16(2):117–128, 1971.

[BRO65] R. Brockett. Optimization theory and the converse of the circle criterion. In *National Electronics Conference*, volume 21, pages 697–701, San Francisco, California, 1965.

[BRO66] R. W. Brockett. The status of stability theory for deterministic systems. *IEEE Trans. Aut. Control*, AC-11:596–606, July 1966.

[BRO70] R. W. Brockett. *Finite Dimensional Linear Systems.* John Wiley & Sons, New York, 1970.

[BRO77] R. Brockett. On improving the circle criterion. In *Proc. IEEE Conf. on Decision and Control*, pages 255–257, 1977.

[BT79] R. K. Brayton and C. H. Tong. Stability of dynamical systems: A constructive approach. *IEEE Trans. Circuits Syst.*, CAS-26(4):224–234, 1979.

[BT80] R. K. Brayton and C. H. Tong. Constructive stability and asymptotic stability of dynamical systems. *IEEE Trans. Circuits Syst.*, CAS-27(11):1121–1130, 1980.

[BW65] R. W. Brockett and J. L. Willems. Frequency domain stability criteria. *IEEE Trans. Aut. Control*, AC-10:255–261, 401–413, 1965.

[BW92] M. Bakonyi and H. J. Woerdeman. The central method for positive semi-definite, contractive and strong Parrott type completion problems. *Operator Theory: Advances and Applications*, 59:78–95, 1992.

[BY89] S. Boyd and Q. Yang. Structured and simultaneous Lyapunov functions for system stability problems. *Int. J. Control*, 49(6):2215–2240, 1989.

[BYDM93] R. D. Braatz, P. M. Young, J. C. Doyle, and M. Morari. Computational complexity of μ calculation. In *Proc. American Control Conf.*, pages 1682–1683, San Francisco, CA, June 1993.

[CD91] G. E. Coxson and C. L. DeMarco. Testing robust stability of general matrix polytopes is an NP-hard computation. In *Proc. Annual Allerton Conf. on Communication, Control and Computing*, pages 105–106, Allerton House, Monticello, Illinois, 1991.

[CD92] G. E. Coxson and C. L. DeMarco. Computing the real structured singular value is NP-hard. Technical Report ECE-92-4, Dept. of Elec. and Comp. Eng., Univ. of Wisconsin-Madison, June 1992.

[CDW75] J. Cullum, W. Donath, and P. Wolfe. The minimization of certain nondifferentiable sums of eigenvalues of symmetric matrices. *Math. Programming Study*, 3:35–55, 1975.

[CFA90] H.-D. Chiang and L. Fekih-Ahmed. A constructive methodology for estimating the stability regions of interconnected nonlinear systems. *IEEE Trans. Circuits Syst.*, 37(5):577–588, 1990.

[CH93] E. B. Castelan and J. C. Hennet. On invariant polyhedra of continuous-time linear systems. *IEEE Trans. Aut. Control*, 38(11):1680–1685, December 1993.

[CHD93] E. G. Collins, W. M. Haddad, and L. D. Davis. Riccati equation approaches for robust stability and performance. In *Proc. American Control Conf.*, volume 2, pages 1079–1083, San Francisco, California, June 1993.

[CHE80A] F. L. Chernousko. Optimal guaranteed estimates of indeterminancies with the aid of ellipsoids I. *Engineering Cybernetics*, 18(3):1–9, 1980.

[CHE80B] F. L. Chernousko. Optimal guaranteed estimates of indeterminancies with the aid of ellipsoids II. *Engineering Cybernetics*, 18(4):1–9, 1980.

[CHE80C] F. L. Chernousko. Optimal guaranteed estimates of indeterminancies with the aid of ellipsoids III. *Engineering Cybernetics*, 18(5):1–7, 1980.

[CL90] M. Corless and G. Leitmann. Deterministic control of uncertain systems: a Lyapunov theory approach. In A. Zinober, editor, *Deterministic control of uncertain systems.* Peregrinus Ltd., 1990.

[CM81] B. Craven and B. Mond. Linear programming with matrix variables. *Linear Algebra and Appl.*, 38:73–80, 1981.

[CNF92] J. Chen, C. N. Nett, and M. K. H. Fan. Worst-case identification in $\mathbf{H}_\infty$: Validation of a priori information, essentially optimal algorithms, and error bounds. In *Proc. American Control Conf.*, pages 251–257, Chicago, June 1992.

[COR85] M. Corless. Stabilization of uncertain discrete-time systems. *Proceedings of the IFAC workshop on model error concepts and compensation, Boston, USA*, pages 125–128, 1985.

[COR90] M. Corless. Guaranteed rates of exponential convergence for uncertain dynamical systems. *Journal of Optimization Theory and Applications*, 64:481–494, 1990.

[COR94] M. Corless. Robust stability analysis and controller design with quadratic Lyapunov functions. In A. Zinober, editor, *Variable structure and Lyapunov control*, volume 193 of *Lecture notes in control and information sciences*, chapter 9, pages 181–203. Springer-Verlag, 1994.

[CP72] S. Chang and T. Peng. Adaptive guaranteed cost control of systems with uncertain parameters. *IEEE Trans. Aut. Control*, AC-17:474–483, August 1972.

[CP84] B. C. Chang and J. B. Pearson, Jr. Optimal disturbance rejection in linear multivariable systems. *IEEE Trans. Aut. Control*, AC-29(10):880–887, October 1984.

[CPS92] R. Cottle, J. S. Pang, and R. E. Stone. *The Linear Complementarity Problem.* Academic Press, 1992.

[CS92A] R. Chiang and M. G. Safonov. *Robust Control Toolbox.* The Mathworks, Inc., 1992.

[CS92B] R. Y. Chiang and M. G. Safonov. Real K_m-synthesis via generalized Popov multipliers. In *Proc. American Control Conf.*, volume 3, pages 2417–2418, Chicago, June 1992.

[CYP93] M. Cheung, S. Yurkovich, and K. M. Passino. An optimal volume ellipsoid algorithm for parameter estimation. *IEEE Trans. Aut. Control*, AC-38(8):1292–1296, 1993.

[DAN92] J. Dancis. Positive semidefinite completions of partial hermitian matrices. *Linear Algebra and Appl.*, 175:97–114, 1992.

[DAN93] J. Dancis. Choosing the inertias for completions of certain partially specified matrices. *SIAM Jour. Matrix Anal. Appl.*, 14(3):813–829, July 1993.

[DAV94] J. David. *Algorithms for analysis and design of robust controllers.* PhD thesis, Kat. Univ. Leuven, ESAT, 3001 Leuven, Belgium, 1994.

[DEL89] J. Deller. Set membership identification in digital signal processing. *IEEE Trans. Acoust., Speech, Signal Processing*, 6(4):4–20, 1989.

[DES65] C. A. Desoer. A generalization of the Popov criterion. *IEEE Trans. Aut. Control*, AC-10:182–185, 1965.

[DG81] H. Dym and I. Gohberg. Extensions of band matrices with band inverses. *Linear Algebra and Appl.*, 36:1–24, 1981.

[DGK79] P. Delsarte, Y. Genin, and Y. Kamp. The Nevanlinna-Pick problem for matrix-valued functions. *SIAM J. Appl. Math.*, 36(1):47–61, February 1979.

[DGK81] P. Delsarte, Y. Genin, and Y. Kamp. On the role of the Nevanlinna-Pick problem in circuit and systems theory. *Circuit Theory and Appl.*, 9:177–187, 1981.

[DIK67] I. Dikin. Iterative solution of problems of linear and quadratic programming. *Soviet Math. Dokl.*, 8(3):674–675, 1967.

[DK71] E. J. Davison and E. M. Kurak. A computational method for determining quadratic Lyapunov functions for non-linear systems. *Automatica*, 7(5):627–636, September 1971.

[DKW82] C. Davis, W. M. Kahan, and H. F. Weinberger. Norm-preserving dilations and their applications to optimal error bounds. *SIAM J. Numerical Anal.*, 19(3):445–469, June 1982.

[DOO81] P. Van Dooren. A generalized eigenvalue approach for solving Riccati equations. *SIAM J. Sci. Stat. Comp.*, 2:121–135, June 1981.

[DOY82] J. Doyle. Analysis of feedback systems with structured uncertainties. *IEE Proc.*, 129-D(6):242–250, November 1982.

[DRT92] D. Den Hertog, C. Roos, and T. Terlaky. A large-step analytic center method for a class of smooth convex programming problems. *SIAM J. Optimization*, 2(1):55–70, February 1992.

[DSFX93] C. E. de Souza, M. Fu, and L. Xie. $\mathbf{H}_\infty$ analysis and synthesis of discrete-time systems with time-varying uncertainty. *IEEE Trans. Aut. Control*, AC-38(3), March 1993.

[DV75] C. A. Desoer and M. Vidyasagar. *Feedback Systems: Input-Output Properties.* Academic Press, New York, 1975.

[EBFB92] L. El Ghaoui, V. Balakrishnan, E. Feron, and S. Boyd. On maximizing a robustness measure for structured nonlinear perturbations. In *Proc. American Control Conf.*, volume 4, pages 2923–2924, Chicago, June 1992.

[EG93] L. El Ghaoui and P. Gahinet. Rank minimization under LMI constraints: a framework for output feedback problems. In *Proc. of the European Contr. Conf.*, pages 1176–1179, July 1993.

[EG94] L. El Ghaoui. Analysis and state-feedback synthesis for systems with multiplicative noise. *Syst. Control Letters*, 1994. Submitted.

[EHM92] L. Elsner, C. He, and V. Mehrmann. Minimization of the norm, the norm of the inverse and the condition number of a matrix by completion. Technical Report 92-028, Universität Bielefeld, POB 8640, D-4800 Bielefeld 1, Germany, 1992.

[EHM93] L. Elsner, C. He, and V. Mehrmann. Minimization of the condition number of a positive-definite matrix by completion. Technical report, TU-Chemnitz-Zwikau, Chemnitz, Germany, September 1993.

[EP92] A. El Bouhtouri and A. J. Pritchard. Stability radii of linear systems with respect to stochastic perturbations. *Syst. Control Letters*, 19(1):29–33, July 1992.

[EP94] A. El Bouthouri and A. Pritchard. A Riccati equation approach to maximizing the stability radius of a linear system under structured stochastic Lipschitzian perturbations. *Syst. Control Letters*, January 1994.

[EZKN90] S. V. Emelyanov, P. V. Zhivoglyadov, S. K. Korovin, and S. V. Nikitin. Group of isometries and some stabilization and stabilizability aspects of uncertain systems. *Problems of Control and Information Theory*, 19(5-6):451–477, 1990.

[FAN93] M. K. H. Fan. A quadratically convergent algorithm on minimizing the largest eigenvalue of a symmetric matrix. *Linear Algebra and Appl.*, 188-189:231–253, 1993.

[FAU67] P. Faurre. *Representation of Stochastic Processes.* PhD thesis, Stanford University, 1967.

[FAU73] P. Faurre. Réalisations Markoviennes de processus stationnaires. Technical Report 13, INRIA, Domaine de Voluceau, Rocquencourt, 78150, Le Chesnay, France, 1973.

[FBB92] E. Feron, V. Balakrishnan, and S. Boyd. Design of stabilizing state feedback for delay systems via convex optimization. In *Proc. IEEE Conf. on Decision and Control*, volume 4, pages 2195–2196, Tucson, December 1992.

[FBBE92] E. Feron, V. Balakrishnan, S. Boyd, and L. El Ghaoui. Numerical methods for $\mathbf{H}_2$ related problems. In *Proc. American Control Conf.*, volume 4, pages 2921–2922, Chicago, June 1992.

[FD77] P. Faurre and M. Depeyrot. *Elements of System Theory.* North-Holland, New York, 1977.

[FED75] P. I. Fedčina. Descriptions of solutions of the tangential Nevanlinna-Pick problem. *Doklady Akad. Nauk. Arm.*, 60:37–42, 1975.

[FER93] E. Feron. *Linear Matrix Inequalities for the Problem of Absolute Stability of Control Systems.* PhD thesis, Stanford University, Stanford, CA 94305, October 1993.

[FG88] M. E. Fisher and J. E. Gayek. Estimating reachable sets for two-dimensional linear discrete systems. *Journal of Optimization Theory and Applications*, 56(1):67–88, January 1988.

[FH82] E. Fogel and Y. Huang. On the value of information in system identification-bounded noise case. *Automatica*, 18(2):229–238, 1982.

[FIL88] A. F. Filippov. *Differential Equations with Discontinuous Righthand Sides.* Kluwer Academic Publishers, Dordrect, The Netherlands, 1988. Translated from 1960 Russian version.

[FIN37] P. Finsler. Über das Vorkommen definiter und semi-definiter Formen in Scharen quadratischer Formen. *Comentarii Mathematici Helvetici*, 9:199–192, 1937.

[FLE85] R. Fletcher. Semidefinite matrix constraints in optimization. *SIAM J. Control and Opt.*, 23:493–513, 1985.

[FM68] A. Fiacco and G. McCormick. *Nonlinear programming: sequential unconstrained minimization techniques.* Wiley, 1968. Reprinted 1990 in the SIAM Classics in Applied Mathematics series.

[FN84] T. Fujii and M. Narazaki. A complete optimality condition in the inverse problem of optimal control. *SIAM J. on Control and Optimization*, 22(2):327–341, March 1984.

[FN92] M. K. H. Fan and B. Nekooie. On minimizing the largest eigenvalue of a symmetric matrix. In *Proc. IEEE Conf. on Decision and Control*, pages 134–139, December 1992.

[FNO87] S. Friedland, J. Nocedal, and M. Overton. The formulation and analysis of numerical methods for inverse eigenvalue problems. *SIAM J. Numer. Anal.*, 24(3):634–667, June 1987.

[FOG79] E. Fogel. System identification via membership set constraints with energy constrained noise. *IEEE Trans. Aut. Control*, AC-24(5):752–757, 1979.

[FP62] M. Fiedler and V. Ptak. On matrices with non-positive off-diagonal elements and positive principal minors. *Czechoslovak Mathematical Journal*, 87:382–400, 1962.

[FP66] M. Fiedler and V. Ptak. Some generalizations of positive definiteness and monotonicity. *Numerische Mathematik*, 9:163–172, 1966.

[FP67] M. Fiedler and V. Ptak. Diagonally dominant matrices. *Czechoslovak Mathematical Journal*, 92:420–433, 1967.

[FR75] W. H. Fleming and R. W. Rishel. *Deterministic and stochastic optimal control.* Springer-Verlag, 1975.

[FRA87] B. A. Francis. *A course in* $\mathbf{H}_\infty$ *Control Theory*, volume 88 of *Lecture Notes in Control and Information Sciences.* Springer-Verlag, 1987.

[FS55] G. E. Forsythe and E. G. Straus. On best conditioned matrices. *Proc. Amer. Math. Soc.*, 6:340–345, 1955.

[FT86] M. K. H. Fan and A. L. Tits. Characterization and efficient computation of the structured singular value. *IEEE Trans. Aut. Control*, AC-31(8):734–743, August 1986.

[FT88] M. K. H. Fan and A. L. Tits. m-form numerical range and the computation of the structured singular value. *IEEE Trans. Aut. Control*, AC-33(3):284–289, March 1988.

[FTD91] M. K. H. Fan, A. L. Tits, and J. C. Doyle. Robustness in the presence of mixed parametric uncertainty and unmodeled dynamics. *IEEE Trans. Aut. Control*, AC-36(1):25–38, January 1991.

[FY79] A. L. Fradkov and V. A. Yakubovich. The $\mathcal{S}$-procedure and duality relations in nonconvex problems of quadratic programming. *Vestnik Leningrad Univ. Math.*, 6(1):101–109, 1979. In Russian, 1973.

[GA93] P. Gahinet and P. Apkarian. A LMI-based parametrization of all $\mathbf{H}_\infty$ controllers with applications. In *Proc. IEEE Conf. on Decision and Control*, pages 656–661, San Antonio, Texas, December 1993.

[GA94] P. Gahinet and P. Apkarian. A convex characterization of parameter dependent $\mathbf{H}_\infty$ controllers. *IEEE Trans. Aut. Control*, 1994. To appear.

[GAH92] P. Gahinet. A convex parametrization of $\mathbf{H}_\infty$ suboptimal controllers. In *Proc. IEEE Conf. on Decision and Control*, pages 937–942, 1992.

[GAY91] J. E. Gayek. A survey of techniques for approximating reachable and controllable sets. In *Proc. IEEE Conf. on Decision and Control*, pages 1724–1729, Brighton, December 1991.

[GB86] A. R. Galimidi and B. R. Barmish. The constrained Lyapunov problem and its application to robust output feedback stabilization. *IEEE Trans. Aut. Control*, AC-31(5):410–419, May 1986.

[GCG93] F. Garofalo, G. Celentano, and L. Glielmo. Stability robustness of interval matrices via Lyapunov quadratic forms. *IEEE Trans. Aut. Control*, 38(2):281–284, February 1993.

[GCZS93] K. M. Grigoriadis, M. J. Carpenter, G. Zhu, and R. E. Skelton. Optimal redesign of linear systems. In *Proc. American Control Conf.*, volume 3, pages 2680–2684, San Francisco, California, June 1993.

[GDSS93] J. C. Geromel, C. C. de Souza, and R. E. Skelton. Static output feedback controllers: Robust stability and convexity. *Submitted to IEEE Trans. Aut. Cont.*, 1993.

[GER85] J. C. Geromel. On the determination of a diagonal solution of the Lyapunov equation. *IEEE Trans. Aut. Control*, AC-30:404–406, April 1985.

[GFS93] K. M. Grigoriadis, A. E. Frazho, and R. E. Skelton. Application of alternating convex projections methods for computation of positive Toeplitz matrices. *IEEE Trans. Signal Proc.*, September 1993. To appear.

[GG94] P. B. Gapski and J. C. Geromel. A convex approach to the absolute stability problem. *IEEE Trans. Aut. Control*, November 1994.

[GHA90] L. El Ghaoui. *Robustness of Linear Systems against Parameter Variations.* PhD thesis, Stanford University, Stanford CA 94305, March 1990.

[GHA94] L. El Ghaoui. Quadratic Lyapunov functions for rational systems: An LMI approach. Submitted to *IFAC Conf. on Robust Control.*, 1994.

[GIL66] E. Gilbert. An iterative procedure for computing the minimum of a quadratic form on a convex set. *SIAM J. on Control*, 4(1):61–80, 1966.

[GJSW84] R. Grone, C. R. Johnson, E. M Sá, and H. Wolkowicz. Positive definite completions of partial Hermitian matrices. *Linear Algebra and Appl.*, 58:109–124, 1984.

[GL81] P. Gács and L. Lovász. Khachiyan's algorithm for linear programming. *Math. Program. Studies*, 14:61–68, 1981.

[GL89] G. Golub and C. Van Loan. *Matrix Computations.* Johns Hopkins Univ. Press, Baltimore, second edition, 1989.

[GL93A] M. Gevers and G. Li. *Parametrizations in Control, Estimation and Filtering Problems: Accuracy Aspects.* Communications and Control Engineering. Springer-Verlag, London, January 1993.

[GL93B] K. Gu and N. K. Loh. Direct computation of stability bounds for systems with polytopic uncertainties. *IEEE Trans. Aut. Control*, 38(2):363–366, February 1993.

[GLS88] M. Grötschel, L. Lovász, and A. Schrijver. *Geometric Algorithms and Combinatorial Optimization*, volume 2 of *Algorithms and Combinatorics.* Springer-Verlag, 1988.

[GN93] P. Gahinet and A. Nemirovskii. *LMI Lab: A Package for Manipulating and Solving LMIs.* INRIA, 1993.

[GP82] S. Gutman and Z. Palmor. Properties of min-max controllers in uncertain dynamical systems. *SIAM Journal on Control and Optimization*, 20(6):850–861, November 1982.

[GPB91] J. C. Geromel, P. L. D. Peres, and J. Bernussou. On a convex parameter space method for linear control design of uncertain systems. *SIAM J. Control and Optimization*, 29(2):381–402, March 1991.

[GR88] D. P. Goodall and E. P. Ryan. Feedback controlled differential inclusions and stabilization of uncertain dynamical systems. *SIAM Journal on Control and Optimization*, 26(6):1431–1441, November 1988.

[GS88] R. R. E. De Gaston and M. G. Safonov. Exact calculation of the multiloop stability margin. *IEEE Trans. Aut. Control*, AC-33(2):156–171, 1988.

[GS92] K. M. Grigoriadis and R. E. Skelton. Alternating convex projection methods for covariance control design. In *Proc. Annual Allerton Conf. on Communication, Control and Computing*, Allerton House, Monticello, Illinois, 1992.

[GT88] C. Goh and D. Teo. On minimax eigenvalue problems via constrained optimization. *Journal of Optimization Theory and Applications*, 57(1):59–68, 1988.

[GTS+94] K. C. Goh, L. Turan, M. G. Safonov, G. P. Papavassilopoulos, and J. H. Ly. Biaffine matrix inequality properties and computational methods. In *Proc. American Control Conf.*, 1994.

[GU92A] K. Gu. Designing stabilizing control of uncertain systems by quasiconvex optimization. *IEEE Trans. Aut. Control*, AC-38(9), 1992.

[GU92B] K. Gu. Linear feedback stabilization of arbitrary time-varying uncertain systems. In *Proc. American Control Conf.*, Chicago, June 1992.

[GU93] K. Gu. The trade-off between $\mathbf{H}_\infty$ attenuation and uncertainty. In *Proc. American Control Conf.*, pages 1666–1670, San Fransico, June 1993.

[GU94] K. Gu. Designing stabilizing control of uncertain systems by quasiconvex optimization. *IEEE Trans. Aut. Control*, 39(1):127–131, January 1994.

[GUT79] S. Gutman. Uncertain dynamical systems — a Lyapunov min-max approach. *IEEE Trans. Aut. Control*, AC-24(3):437–443, June 1979.

[GV74] G. Golub and J. Varah. On a characterization of the best ℓ_2 scaling of a matrix. *SIAM J. on Numerical Analysis*, 11(3):472–479, June 1974.

[GZL90] K. Gu, M. A. Zohdy, and N. K. Loh. Necessary and sufficient conditions of quadratic stability of uncertain linear systems. *IEEE Trans. Aut. Control*, AC-35(5):601–604, May 1990.

[HB76] H. P. Horisberger and P. R. Bélanger. Regulators for linear, time invariant plants with uncertain parameters. *IEEE Trans. Aut. Control*, AC-21:705–708, 1976.

[HB80] C. V. Hollot and B. R. Barmish. Optimal quadratic stabilizability of uncertain linear systems. *Proc. 18th Allerton Conf. on Communication, Control and Computing, University of Illinois, Monticello*, pages 697–706, 1980.

[HB91A] W. Haddad and D. Bernstein. Parameter-dependent Lyapunov functions, constant real parameter uncertainty, and the Popov criterion in robust analysis and synthesis. In *Proc. IEEE Conf. on Decision and Control*, pages 2274–2279, 2632–2633, Brighton, U.K., December 1991.

[HB91B] J. Hennet and J. Beziat. A class of invariant regulators for the discrete-time linear constrained regulator problem. *Automatica*, 27:549–554, 1991.

[HB93A] W. Haddad and D. Bernstein. Explicit construction of quadratic Lyapunov functions for the small gain, positivity, circle, and Popov theorems and their application to robust stability. part I: Continuous-time theory. *Int. J. Robust and Nonlinear Control*, 3:313–339, 1993.

[HB93B] W. Haddad and D. Bernstein. Off-axis absolute stability criteria and μ-bounds involving non-positive-real plant-dependent multipliers for robust stability and performance with locally slope-restricted monotonic nonlinearities. In *Proc. American Control Conf.*, volume 3, pages 2790–2794, San Francisco, California, June 1993.

[HCB93] W. Haddad, E. Collins, and D. Bernstein. Robust stability analysis using the small gain, circle, positivity, and Popov theorems: A comparative study. *IEEE Trans. Control Sys. Tech.*, 1(4):290–293, December 1993.

[HCM93] W. M. Haddad, E. G. Collins, and R. Moser. Fixed structure computation of the structured singular value. In *Proc. American Control Conf.*, pages 1673–1674, San Fransico, June 1993.

[HES81] M. R. Hestenes. *Optimization Theory, the Finite Dimensional Case.* Krieger, Huntington, New York, 1981.

[HH93A] S. R. Hall and J. P. How. Mixed $\mathbf{H}_2/\mu$ performance bounds using dissipation theory. In *Proc. IEEE Conf. on Decision and Control*, San Antonio, Texas, December 1993.

[HH93B] J. P. How and S. R. Hall. Connections between the Popov stability criterion and bounds for real parametric uncertainty. In *Proc. American Control Conf.*, volume 2, pages 1084–1089, June 1993.

[HHH93A] J. P. How, W. M. Haddad, and S. R. Hall. Robust control synthesis examples with real parameter uncertainty using the Popov criterion. In *Proc. American Control Conf.*, volume 2, pages 1090–1095, San Francisco, California, June 1993.

[HHH93B] J. P. How, S. R. Hall, and W. M. Haddad. Robust controllers for the Middeck Active Control Experiment using Popov controller synthesis. In *AIAA Guid., Nav., and Control Conf.*, pages 1611–1621, August 1993. Submitted to *IEEE Transactions on Control Systems Technology.*

[HHHB92] W. Haddad, J. How, S. Hall, and D. Bernstein. Extensions of mixed-μ bounds to monotonic and odd monotonic nonlinearities using absolute stability theory. In *Proc. IEEE Conf. on Decision and Control*, Tucson, Arizona, December 1992.

[HJ91] R. Horn and C. Johnson. *Topics in Matrix Analysis.* Cambridge University Press, Cambridge, 1991.

[HOL87] C. V. Hollot. Bound invariant Lyapunov functions: A means for enlarging the class of stabilizable uncertain systems. *Int. J. Control*, 46(1):161–184, 1987.

[HOW93] J. P. How. *Robust Control Design with Real Parameter Uncertainty using Absolute Stability Theory.* PhD thesis, Massachusetts Institute of Technology, February 1993.

[HU87] H. Hu. An algorithm for rescaling a matrix positive definite. *Linear Algebra and its Applications*, 96:131–147, 1987.

[HUA67] P. Huard. *Resolution of mathematical programming with nonlinear constraints by the method of centers.* North Holland, Amsterdam, 1967.

[HUL93] J.-B. Hiriart-Urruty and C. Lemaréchal. *Convex Analysis and Minimization Algorithms I*, volume 305 of *Grundlehren der mathematischen Wissenschaften.* Springer-Verlag, New York, 1993.

[HUY92] J.-B. Hiriart-Urruty and D. Ye. Sensitivity analysis of all eigenvalues of a symmetric matrix. Technical report, Univ. Paul Sabatier, Toulouse, 1992.

[HW93] J. W. Helton and H. J. Woerdeman. Symmetric Hankel operators: Minimal norm extensions and eigenstructures. *Linear Algebra and Appl.*, 185:1–19, 1993.

[IR64] E. S. Ibrahim and Z. V. Rekasius. Construction of Liapunov functions for nonlinear feedback systems with several nonlinear elements. In *Proc. Joint American Control Conf.*, pages 256–260, June 1964.

[IS93A] T. Iwasaki and R. E. Skelton. A complete solution to the general $\mathbf{H}_\infty$ control problem: LMI existence conditions and state space formulas. In *Proc. American Control Conf.*, pages 605–609, San Francisco, June 1993.

[IS93B] T. Iwasaki and R. E. Skelton. Parametrization of all stabilizing controllers via quadratic Lyapunov functions. *Submitted to J. Optim. Theory Appl.*, 1993.

[ISG93] T. Iwasaki, R. E. Skelton, and J. C. Geromel. Linear quadratic suboptimal control via static output feedback. *Submitted to Syst. and Cont. Letters*, 1993.

[ITO51] K. Ito. On a formula concerning stochastic differentials. *Nagoya Math. J.*, 3(55), 1951.

[IWA93] T. Iwasaki. *A Unified Matrix Inequality Approach to Linear Control Design.* PhD thesis, Purdue University, West Lafayette, IN 47907, December 1993.

[JAR91] F. Jarre. Interior-point methods for convex programming. *To appear, Appl. Math. Opt.*, 1991.

[JAR93A] F. Jarre. An interior-point method for convex fractional programming. *Submitted to Math. Progr.*, March 1993.

[JAR93B] F. Jarre. An interior-point method for minimizing the maximum eigenvalue of a linear combination of matrices. *SIAM J. Control and Opt.*, 31(5):1360–1377, September 1993.

[JAR93C] F. Jarre. Optimal ellipsoidal approximations around the analytic center. Technical report, Intitüt für Angewandte Mathematik, University of Würzburg, 8700 Würzburg, Germany, 1993.

[JL65] E. I. Jury and B. W. Lee. The absolute stability of systems with many nonlinearities. *Automation and Remote Control*, 26:943–961, 1965.

[JL92] C. R. Johnson and M. Lundquist. An inertia formula for Hermitian matrices with sparse inverses. *Linear Algebra and Appl.*, 162:541–556, 1992.

[JOH85] F. John. Extremum problems with inequalities as subsidiary conditions. In J. Moser, editor, *Fritz John, Collected Papers*, pages 543–560. Birkhauser, Boston, Massachussetts, 1985.

[KAH68] W. Kahan. Circumscribing an ellipsoid about the intersection of two ellipsoids. *Canad. Math. Bull.*, 11(3):437–441, 1968.

[KAI80] T. Kailath. *Linear Systems*. Prentice-Hall, New Jersey, 1980.

[KAL63A] R. E. Kalman. Lyapunov functions for the problem of Lur'e in automatic control. *Proc. Nat. Acad. Sci., USA*, 49:201–205, 1963.

[KAL63B] R. E. Kalman. On a new characterization of linear passive systems. In *Proc. First Annual Allerton Conf. on Communication, Control and Computing*, pages 456–470, 1963.

[KAL64] R. E. Kalman. When is a linear control system optimal? *Trans. ASME, J. Basic Eng., Ser. D.*, 86:51–60, 1964.

[KAM83] V. A. Kamenetskii. Absolute stability and absolute instability of control systems with several nonlinear nonstationary elements. *Automation and Remote Control*, 44(12):1543–1552, 1983.

[KAM89] V. A. Kamenetskii. Convolution method for matrix inequalities. *Automation and Remote Control*, 50(5):598–607, 1989.

[KAR84] N. Karmarkar. A new polynomial-time algorithm for linear programming. *Combinatorica*, 4(4):373–395, 1984.

[KAV94] D. Kavranoğlu. Computation of the optimal solution for the zeroth order $\mathbf{L}_\infty$ norm approximation problem. *Syst. Control Letters*, 1994. Submitted.

[KB93] E. Kaszkurewicz and A. Bhaya. Robust stability and diagonal Lyapunov functions. *SIAM J. on Matrix Analysis and Applications*, 14(2):508–520, April 1993.

[KHA78] V. L. Kharitonov. Asymptotic stability of an equilibrium position of a family of systems of linear differential equations. *Differential'nye Uraveniya*, 14(11):1483–1485, 1978.

[KHA79] L. Khachiyan. A polynomial algorithm in linear programming. *Soviet Math. Dokl.*, 20:191–194, 1979.

[KHA82] H. K. Khalil. On the existence of positive diagonal P such that $PA + A^T P < 0$. *IEEE Trans. Aut. Control*, AC-27:181–184, 1982.

[KIM87] H. Kimura. Directional interpolation approach to $\mathbf{H}^\infty$-optimization and robust stabilization. *IEEE Trans. Aut. Control*, 32(12):1085–1093, December 1987.

[KIS91] M. Kisielewicz. *Differential inclusions and optimal control*, volume 44 of *Mathematics and its applications; East European series*. Kluwer Academic Publishers, Boston, 1991.

[KK60] I. I. Kats and N. N. Krasovskii. On the stability of systems with random parameters. *Prikl. Mat. Mek.*, 809, 1960.

[KK92] L. Khachiyan and B. Kalantari. Diagonal matrix scaling and linear programming. *SIAM J. on Optimization*, 2(4):668–672, November 1992.

[KKR93] I. Kaminer, P. P. Khargonekar, and M. A. Rotea. Mixed $\mathbf{H}_2/\mathbf{H}_\infty$ control for discrete-time systems via convex optimization. *Automatica*, 29(1):57–70, 1993.

[KL85A] B. Kouvaritakis and D. Latchman. Necessary and sufficient stability criterion for systems with structured uncertainties: The major principal direction alignment principle. *Int. J. Control*, 42(3):575–598, 1985.

[KL85B] B. Kouvaritakis and D. Latchman. Singular value and eigenvalue techniques in the analysis of systems with structured perturbations. *Int. J. Control*, 41(6):1381–1412, 1985.

[KLB90] R. Kosut, M. Lau, and S. Boyd. Identification of systems with parametric and nonparametric uncertainty. In *Proc. American Control Conf.*, pages 2412–2417, San Diego, May 1990.

[KLB92] R. Kosut, M. Lau, and S. Boyd. Set-membership identification of systems with parametric and nonparametric uncertainty. *IEEE Trans. Aut. Control*, AC-37(7):929–941, July 1992.

[KLE69] D. L. Kleinman. On the stability of linear stochastic systems. *IEEE Trans. Aut. Control*, August 1969.

[KOK92] P. V. Kokotovic. The joy of feedback: Nonlinear and adaptive. *IEEE Control Systems*, June 1992.

[KOZ69] F. Kozin. A survey of stability of stochastic systems. *Automatica*, 5:95–112, 1969.

[KP87A] V. A. Kamenetskii and E. S. Pyatnitskii. Gradient method of constructing Lyapunov functions in problems of absolute stability. *Automation and Remote Control*, pages 1–9, 1987.

[KP87B] V. A. Kamenetskii and Y. S. Pyatnitskii. An iterative method of Lyapunov function construction for differential inclusions. *Syst. Control Letters*, 8:445–451, 1987.

[KPR88] P. P. Khargonekar, I. R. Petersen, and M. A. Rotea. $\mathbf{H}_\infty$-optimal control with state-feedback. *IEEE Trans. Aut. Control*, AC-33(8):786–788, August 1988.

[KPY91] K. Kortanek, F. Potra, and Y. Ye. On some efficient interior point methods for nonlinear convex programming. *Linear Algebra and Appl.*, 152:169–189, 1991.

[KPZ90] P. P. Khargonekar, I. R. Petersen, and K. Zhou. Robust stabilization of uncertain linear systems: quadratic stabilizability and $\mathbf{H}_\infty$ control theory. *IEEE Trans. Aut. Control*, AC-35(3):356–361, March 1990.

[KR88] P. P. Khargonekar and M. A. Rotea. Stabilization of uncertain systems with norm bounded uncertainty using control Lyapunov functions. *Proc. IEEE Conf. on Decision and Control*, 1:503–507, 1988.

[KR91] P. P. Khargonekar and M. A. Rotea. Mixed $\mathbf{H}_2/\mathbf{H}_\infty$ control: A convex optimization approach. *IEEE Trans. Aut. Control*, AC-36(7):824–837, July 1991.

[KRA56] N. N. Krasovskii. Application of Lyapunov's second method for equations with time delay. *Prikl. Mat. Mek.*, 20(3):315–327, 1956.

[KRA91] J. Kraaijevanger. A characterization of Lyapunov diagonal stability using Hadamard products. *Linear Algebra and Appl.*, 151:245–254, 1991.

[KRB89] N. Khraishi, R. Roy, and S. Boyd. A specification language approach to solving convex antenna weight design problems. Technical Report 8901, Stanford University, Information Systems Laboratory, Electrical Engineering Department, January 1989.

[KS75] V. J. S. Karmarkar and D. D. Siljak. Maximization of absolute stability regions by mathematical programming methods. *Regelungstechnik*, 2:59–61, 1975.

[KT93] L. G. Khachiyan and M. J. Todd. On the complexity of approximating the maximal inscribed ellipsoid for a polytope. *Mathematical Programming*, 61:137–159, 1993.

[KUS67] H. J. Kushner. *Stochastic stability and control.* Academic Press, New York, 1967.

[LAU79] A. J. Laub. A Schur method for solving algebraic Riccati equations. *IEEE Trans. Aut. Control*, AC-24(6):913–921, December 1979.

[LEF65] S. Lefschetz. *Stability of Nonlinear Control Systems.* Mathematics in Science and Engineering. Academic Press, New York, 1965.

[LEI79] G. Leitmann. Guaranteed asymptotic stability for some linear systems with bounded uncertainties. *Trans. of the ASME: Journal of Dynamic Systems, Measurement, and Control*, 101:212–216, September 1979.

[LET61] A. M. Letov. *Stability in Nonlinear Control Systems.* Princeton University Press, Princeton, 1961.

[LEV59] J. J. Levin. On the matrix Riccati equation. *Proc. Amer. Math. Soc.*, 10:519–524, 1959.

[LH66] B. Lieu and P. Huard. La méthode des centres dans un espace topologique. *Numerische Mathematik*, 8:56–67, 1966.

[LIU68] R. W. Liu. Convergent systems. *IEEE Trans. Aut. Control*, AC-13(4):384–391, August 1968.

[LKB90] M. Lau, R. Kosut, and S. Boyd. Parameter set estimation of systems with uncertain nonparametric dynamics and disturbances. In *Proc. IEEE Conf. on Decision and Control*, pages 3162–3167, Honolulu, Hawaii, December 1990.

[LLJ90] R. Lozano-Leal and S. Joshi. Strictly positive real transfer functions revisited. *IEEE Trans. Aut. Control*, AC-35:1243–1245, 1990.

[LN92] H. A. Latchman and R. J. Norris. On the equivalence between similarity and nonsimilarity scalings in robustness analysis. *IEEE Trans. Aut. Control*, AC-37(11):1849–1853, 1992.

[LOG90] H. Logemann. Circle criteria, small-gain conditions and internal stability for infinite-dimensional systems. Technical Report 222, Institut für Dynamische Systeme, Universität Bremen, April 1990.

[LP44] A. I. Lur'e and V. N. Postnikov. On the theory of stability of control systems. *Applied mathematics and mechanics*, 8(3), 1944. In Russian.

[LR80] P. Lancaster and L. Rodman. Existence and uniqueness theorems for the algebraic Riccati equation. *Int. J. Control*, 32(2):285–309, 1980.

[LSA94] J. H. Ly, M. G. Safonov, and F. Ahmad. Positive real Parrott theorem with application to LMI controller synthesis. In *Proc. American Control Conf.*, 1994.

[LSC94] J. H. Ly, M. G. Safonov, and R. Y Chiang. Real/complex multivariable stability margin computation via generalized Popov multiplier — LMI approach. In *Proc. American Control Conf.*, 1994.

[LSG92] K. Liu, R. E. Skelton, and K. Grigoriadis. Optimal controllers for finite wordlength implementation. *IEEE Trans. Aut. Control*, 372, 1992.

[LSL69] R. Liu, R. Saeks, and R. J. Leake. On global linearization. *SIAM-AMS proceedings*, pages 93–102, 1969.

[LUR57] A. I. Lur'e. *Some Nonlinear Problems in the Theory of Automatic Control.* H. M. Stationery Off., London, 1957. In Russian, 1951.

[LYA47] A. M. Lyapunov. *Problème général de la stabilité du mouvement*, volume 17 of *Annals of Mathematics Studies.* Princeton University Press, Princeton, 1947.

[LZD91] W. M. Lu, K. Zhou, and J. C. Doyle. Stabilization of LFT systems. In *Proc. IEEE Conf. on Decision and Control*, volume 2, pages 1239–1244, Brighton, December 1991.

[MAK91] P. Mäkilä. Multiple models, multiplicative noise and linear quadratic control-algorithm aspects. *Int. J. Control*, 54(4):921–941, October 1991.

[MAG76] E. F. Mageirou. Values and strategies in infinite duration linear quadratic games. *IEEE Trans. Aut. Control*, AC-21:547–550, 1976.

[MCL71] P. J. McLane. Optimal stochastic control of linear systems with state- and control-dependent disturbances. *IEEE Trans. Aut. Control*, 16(6):292–299, December 1971.

[MEG92A] A. Megretsky. $\mathcal{S}$-procedure in optimal non-stochastic filtering. Technical Report TRITA/MAT-92-0015, Department of Mathematics, Royal Institute of Technology, S-100 44 Stockholm, Sweden, March 1992.

[MEG92B] A. Megretsky. Power distribution approach in robust control. Technical Report TRITA/MAT-92-0027, Department of Mathematics, Royal Institute of Technology, S-100 44 Stockholm, Sweden, 1992.

[MEG93] A. Megretsky. Necessary and sufficient conditions of stability: A multiloop generalization of the circle criterion. *IEEE Trans. Aut. Control*, AC-38(5):753–756, May 1993.

[MEH91] S. Mehrotra. On the implementation of a primal-dual interior point method. *SIAM Journal on Optimization*, 2(4):575–601, November 1991.

[MEI91] A. M. Meilakhs. Construction of a Lyapunov function for parametrically perturbed linear systems. *Automation and Remote Control*, 52(10):1479–1481, October 1991.

[MEY66] K. R. Meyer. On the existence of Lyapunov functions for the problem of Lur'e. *J. SIAM Control*, 3(3):373–383, 1966.

[MH78] P. J. Moylan and D. J. Hill. Stability criteria for large-scale systems. *IEEE Trans. Aut. Control*, AC-23:143–149, 1978.

[MOL87] A. P. Molchanov. Lyapunov functions for nonlinear discrete-time control systems. *Automation and Remote Control*, pages 728–736, 1987.

[MP86] A. P. Molchanov and E. S. Pyatnitskii. Lyapunov functions that specify necessary and sufficient conditions of absolute stability of nonlinear nonstationary control system Pt. I–III. *Automation and Remote Control*, 46(3,4,5):(344–354), (443–451), (620–630), 1986.

[MP89] A. P. Molchanov and E. S. Pyatnitskii. Criteria of asymptotic stability of differential and difference inclusions encountered in control theory. *Syst. Control Letters*, 13:59–64, 1989.

[MR76] C. T. Mullis and R. A. Roberts. Synthesis of minimum roundoff noise fixed-point digital filters. *IEEE Trans. Circuits Syst.*, 23:551–562, September 1976.

[MV63] S. G. Margolis and W. G. Vogt. Control engineering applications of V. I. Zubov's construction procedure for Lyapunov functions. *IEEE Trans. Aut. Control*, AC-8:104–113, April 1963.

[NEM94] A. Nemirovskii. Several NP-Hard problems arising in robust stability analysis. *Mathematics of Control, Signals, and Systems*, 6(1):99–105, 1994.

[NEV19] R. Nevanlinna. Über beschränkte funktionen die in gegebenen punkten vorgeschreibene funktionswerte bewirkt werden. *Anal. Acad. Sci. Fenn., Ser. A*, 13:1–71, 1919.

[NF92] B. Nekooie and M. K. H. Fan. A quadratically convergent local algorithm on minimizing sums of the largest eigenvalues of a symmetric matrix. *Submitted to Computational Optimization and Applications.*, October 1992.

[NN88] Yu. Nesterov and A. Nemirovsky. A general approach to polynomial-time algorithms design for convex programming. Technical report, Centr. Econ. & Math. Inst., USSR Acad. Sci., Moscow, USSR, 1988.

[NN90A] Yu. Nesterov and A. Nemirovsky. *Optimization over positive semidefinite matrices: Mathematical background and user's manual.* USSR Acad. Sci. Centr. Econ. & Math. Inst., 32 Krasikova St., Moscow 117418 USSR, 1990.

[NN90B] Yu. Nesterov and A. Nemirovsky. Self-concordant functions and polynomial time methods in convex programming. Technical report, Centr. Econ. & Math. Inst., USSR Acad. Sci., Moscow, USSR, April 1990.

[NN91A] Yu. Nesterov and A. Nemirovsky. Conic formulation of a convex programming problem and duality. Technical report, Centr. Econ. & Math. Inst., USSR Academy of Sciences, Moscow USSR, 1991.

[NN91B] Yu. Nesterov and A. Nemirovsky. An interior point method for generalized linear-fractional programming. *Submitted to Math. Programming, Series B.*, 1991.

[NN93] Yu. E. Nesterov and A. S. Nemirovskii. Acceleration of the path-following method for optimization over the cone of positive semidefinite matrices. Manuscript., 1993.

[NN94] Yu. Nesterov and A. Nemirovsky. *Interior-point polynomial methods in convex programming*, volume 13 of *Studies in Applied Mathematics.* SIAM, Philadelphia, PA, 1994.

[Nor86] J. P. Norton. *An Introduction to Identification.* Academic Press, 1986.

[NS72] T. Nakamizo and Y. Sawaragi. Analytical study on n-th order linear system with stochastic coefficients. In R. F. Curtain, editor, *Lecture notes on Mathematics: Stability of stochastic dynamical systems*, pages 173–185. Springer verlag, 1972.

[NT73] K. S. Narendra and J. H. Taylor. *Frequency domain criteria for absolute stability.* Electrical science. Academic Press, New York, 1973.

[NW92] G. Naevdal and H. J. Woerdeman. Partial matrix contractions and intersections of matrix balls. *Linear Algebra and Appl.*, 175:225–238, 1992.

[NY83] A. Nemirovskii and D. Yudin. *Problem Complexity and Method Efficiency in Optimization.* John Wiley & Sons, 1983.

[OC87] A. I. Ovseevich and F. L. Chernousko. On optimal ellipsoids approximating reachable sets. *Prob. of Control and Info. Th.*, 16(2):125–134, 1987.

[OF85] S. D. O'Young and B. A. Francis. Sensitivity tradeoffs for multivariable plants. *IEEE Trans. Aut. Control*, AC-30(7):625–632, July 1985.

[OKUI88] Y. Ohashi, H. Kando, H. Ukai, and T. Iwazumi. Optimal design of FIR linear phase filters via convex quadratic programming method. *Int. J. Systems Sci.*, 19(8):1469–1482, 1988.

[Ola94] A. Olas. Construction of optimal Lyapunov functions for systems with structured uncertainties. *IEEE Trans. Aut. Control*, 39(1):167–171, January 1994.

[OMS93] E. Ohara, I. Masubuchi, and N. Suda. Quadratic stabilization and robust disturbance attenuation using parametrization of stabilizing state feedback gains – convex optimization approach. In *Proc. American Control Conf.*, volume 3, pages 2816–2820, San Francisco, California, June 1993.

[OS93] A. Olas and S. So. Application of the simplex method to the multi-variable stability problem. In *Proc. American Control Conf.*, pages 1680–1681, San Fransico, June 1993.

[OS94] A. Ohara and T. Sugie. Control systems synthesis via convex optimization. *Systems, Control and Information*, 38(3):139–146, March 1994. Special issue on Numerical Approaches in Control Theory. In Japanese.

[OSA93] A. Ohara, N. Suda, and A. Amari. Dualistic differential geometry of positive definite matrices and its application to optimizations and matrix completion problems. In *MTNS-93*, 1993.

[OV92] D. Obradovic and L. Valavani. Stability and performance in the presence of magnitude bounded real uncertainty: Riccati equation based state space approaches. *Automatica*, July 1992.

[Ove88] M. Overton. On minimizing the maximum eigenvalue of a symmetric matrix. *SIAM J. on Matrix Analysis and Applications*, 9(2):256–268, 1988.

[Ove92] M. Overton. Large-scale optimization of eigenvalues. *SIAM J. Optimization*, pages 88–120, 1992.

[OW92] M. Overton and R. Womersley. On the sum of the largest eigenvalues of a symmetric matrix. *SIAM J. on Matrix Analysis and Applications*, pages 41–45, 1992.

[OW93] M. Overton and R. Womersley. Optimality conditions and duality theory for minimizing sums of the largest eigenvalues of symmetric matrices. *Mathematical Programming*, 62:321–357, 1993.

[Pac94] A. Packard. Gain scheduling via linear fractional transformations. *Syst. Control Letters*, 22:79–92, 1994.

[Pan89] E. Panier. On the need for special purpose algorithms for minimax eigenvalue problems. *Journal Opt. Theory Appl.*, 62(2):279–287, August 1989.

[Par78] S. Parrott. On a quotient norm and the Sz.-Nagy-Foias lifting theorem. *Journal of Functional Analysis*, 30:311–328, 1978.

[PB84] I. Petersen and B. Barmish. The stabilization of single-input uncertain linear systems via control. In *Proc. 6th International Conference on Analysis and Optimization of Systems*, pages 69–83. springer-Verlag, 1984. Lecture notes in control and information sciences, Vol. 62.

[PB92] A. Packard and G. Becker. Quadratic stabilization of parametrically-dependent linear systems using parametrically-dependent linear, dynamic feedback. *ASME J. of Dynamic Systems, Measurement and Control*, July 1992. Submitted.

[PBG89] P. L. D. Peres, J. Bernussou, and J. C. Geromel. A cutting plane technique applied to robust control synthesis. In *IFAC workshop on Control Applications of Nonlinear Programming and Optimization*, Paris, 1989.

[PBPB93] A. Packard, G. Becker, D. Philbrick, and G. Balas. Control of parameter dependent systems: applications to $\mathbf{H}_\infty$ gain scheduling. In *1993 IEEE Regional Conference on Aerospace Control Systems*, pages 329–333, May 1993.

[PD90] A. Packard and J. C. Doyle. Quadratic stability with real and complex perturbations. *IEEE Trans. Aut. Control*, AC-35(2):198–201, February 1990.

[PD93] A. Packard and J. Doyle. The complex structured singular value. *Automatica*, 29(1):71–109, 1993.

[PET85] I. R. Petersen. Quadratic stabilizability of uncertain linear systems: existence of a nonlinear stabilizing control does not imply existence of a stabilizing control. *IEEE Trans. Aut. Control*, AC-30(3):291–293, 1985.

[PET87A] I. R. Petersen. Disturbance attenuation and $\mathbf{H}_\infty$ optimization: a design method based on the algebraic Riccati equation. *IEEE Trans. Aut. Control*, AC-32(5):427–429, May 1987.

[PET87B] I. R. Petersen. A stabilization algorithm for a class of uncertain systems. *Syst. Control Letters*, 8(4):351–357, 1987.

[PET88] I. R. Petersen. Quadratic stabilizability of uncertain linear systems containing both constant and time-varying uncertain parameters. *Journal of Optimization Theory and Applications*, 57(3):439–461, June 1988.

[PET89] I. R. Petersen. Complete results for a class of state-feedback disturbance attenuation problems. *IEEE Trans. Aut. Control*, AC-34(11):1196–1199, November 1989.

[PF92] S. Phoojaruenchanachai and K. Furuta. Memoryless stabilization of uncertain linear systems including time-varying state delays. *IEEE Trans. Aut. Control*, 37(7), July 1992.

[PG93] P. L. D. Peres and J. C. Geromel. $\mathbf{H}_2$ control for discrete-time systems, optimality and robustness. *Automatica*, 29(1):225–228, 1993.

[PGB93] P. L. D. Peres, J. C. Geromel, and J. Bernussou. Quadratic stabilizability of linear uncertain systems in convex-bounded domains. *Automatica*, 29(2):491–493, 1993.

[PGS91] P. L. D. Peres, J. C. Geromel, and S. R. Souza. Convex analysis of discrete-time uncertain $\mathbf{H}_\infty$ control problems. In *Proc. IEEE Conf. on Decision and Control*, Brighton, UK, December 1991.

[PGS93] P. L. D. Peres, J. C. Geromel, and S. R. Souza. $\mathbf{H}_\infty$ robust control by static output feedback. In *Proc. American Control Conf.*, volume 1, pages 620–621, San Francisco, California, June 1993.

[PH86] I. R. Petersen and C. V. Hollot. A Riccati equation approach to the stabilization of uncertain linear systems. *Automatica*, 22(4):397–411, 1986.

[PHI89] Y. A. Phillis. Algorithms for systems with multiplicative noise. *J. Contr. Dyn. Syst.*, 30:65–81, 1989.

[PIC16] G. Pick. Über die beschränkungen analytischer funktionen, welche durch vorgegebene funktionswerte bewirkt werden. *Math. Ann.*, 77:7–23, 1916.

[PKT+94] K. Poolla, P. Khargonekar, A. Tikku, J. Krause, and K. Nagpal. A time-domain approach to model validation. *IEEE Trans. Aut. Control*, May 1994.

[PM92] I. R. Petersen and D. C. McFarlane. Optimal guaranteed cost control of uncertain linear systems. In *Proc. American Control Conf.*, Chicago, June 1992.

[PMR93] I. R. Petersen, D. C. McFarlane, and M. A. Rotea. Optimal guaranteed cost control of discrete-time uncertain systems. In *12th IFAC World Congress*, pages 407–410, Sydney, Australia, July 1993.

[PON61] L. Pontryagin. *Optimal Regulation Processes*, volume 18. American Mathematical Society Translations, 1961.

[PON62] L. Pontryagin. *The Mathematical Theory of Optimal Processes.* John Wiley & Sons, 1962.

[POP62] V. M. Popov. Absolute stability of nonlinear systems of automatic control. *Automation and Remote Control*, 22:857–875, 1962.

[POP64] V. M. Popov. One problem in the theory of absolute stability of controlled systems. *Automation and Remote Control*, 25(9):1129–1134, 1964.

[POP73] V. M. Popov. *Hyperstability of Control Systems.* Springer-Verlag, New York, 1973.

[POS84] M. Post. Minimum spanning ellipsoids. In *Proc. ACM Symp. Theory Comp.*, pages 108–116, 1984.

[POT66] J. E. Potter. Matrix quadratic solutions. *J. SIAM Appl. Math.*, 14(3):496–501, May 1966.

[PP92] G. Picci and S. Pinzoni. On feedback-dissipative systems. *Jour. of Math. Systems, Estimation and Control*, 2(1):1–30, 1992.

[PR91A] E. S. Pyatnitskii and L. B. Rapoport. Existence of periodic motions and test for absolute stability of nonlinear nonstationary systems in the three-dimensional case. *Automation and Remote Control*, 52(5):648–658, 1991.

[PR91B] E. S. Pyatnitskii and L. B. Rapoport. Periodic motion and tests for absolute stability of non-linear nonstationary systems. *Automation and Remote Control*, 52(10):1379–1387, October 1991.

[PR94] S. Poljak and J. Rohn. Checking robust nonsingularity is NP-complete. *Mathematics of Control, Signals, and Systems*, 6(1):1–9, 1994.

[PS82] E. S. Pyatnitskii and V. I. Skorodinskii. Numerical methods of Lyapunov function construction and their application to the absolute stability problem. *Syst. Control Letters*, 2(2):130–135, August 1982.

[PS83] E. S. Pyatnitskii and V. I. Skorodinskii. Numerical method of construction of Lyapunov functions and absolute stability criteria in the form of numerical procedures. *Automation and Remote Control*, 44(11):1427–1437, 1983.

[PS85] F. P. Preparata and M. I. Shamos. *Computational Geometry, an Introduction.* Texts and monographs in Computer Science. Springer-Verlag, New York, 1985.

[PS86] E. S. Pyatnitskii and V. I. Skorodinskii. A criterion of absolute stability of non-linear sampled-data control systems in the form of numerical procedures. *Automation and Remote Control*, 47(9):1190–1198, August 1986.

[PSG92] P. L. D. Peres, S. R. Souza, and J. C. Geromel. Optimal $\mathbf{H}_2$ control for uncertain systems. In *Proc. American Control Conf.*, Chicago, June 1992.

[PW82] E. Polak and Y. Wardi. A nondifferentiable optimization algorithm for the design of control systems subject to singular value inequalities over a frequency range. *Automatica*, 18(3):267–283, 1982.

[PYA68] E. S. Pyatnitskii. New research on the absolute stability of automatic control (review). *Automation and Remote Control*, 29(6):855–881, June 1968.

[PYA70A] E. S. Pyatnitskii. Absolute stability of nonlinear pulse systems with nonstationary nonlinearity. *Automation and Remote Control*, 31(8):1242–1249, August 1970.

[PYA70B] E. S. Pyatnitskii. Absolute stability of nonstationary nonlinear systems. *Automation and Remote Control*, 31(1):1–11, January 1970.

[PYA71] E. S. Pyatnitskii. Criterion for the absolute stability of second-order non-linear controlled systems with one nonlinear nonstationary element. *Automation and Remote Control*, 32(1):1–11, January 1971.

[PZP+92] A. Packard, K. Zhou, P. Pandey, J. Leonhardson, and G. Balas. Optimal, constant I/O similarity scaling for full-information and state-feedback control problems. *Syst. Control Letters*, 19:271–280, October 1992.

[PZPB91] A. Packard, K. Zhou, P. Pandey, and G. Becker. A collection of robust control problems leading to LMI's. In *Proc. IEEE Conf. on Decision and Control*, volume 2, pages 1245–1250, Brighton, December 1991.

[RAP86] L. B. Rapoport. Lyapunov stability and positive definiteness of the quadratic form on the cone. *Prikladnaya matematika i mekhanika*, 50(4):674–679, 1986. In Russian; English translation in *Applied Mathematics and Mechanics*.

[RAP87] L. B. Rapoport. Absolute stability of control systems with several nonlinear stationary elements. *Automation and Remote Control*, pages 623–630, 1987.

[RAP88] L. B. Rapoport. Sign-definiteness of a quadratic form with quadratic constraints and absolute stability of nonlinear control systems. *Sov. Phys. Dokl.*, 33(2):96–98, 1988.

[RAP89] L. B. Rapoport. Frequency criterion of absolute stability for control systems with several nonlinear stationary elements. *Automation and Remote Control*, 50(6):743–750, 1989.

[RAP90] L. B. Rapoport. Boundary of absolute stability of nonlinear nonstationary systems and its connection with the construction of invariant functions. *Automation and Remote Control*, 51(10):1368–1375, 1990.

[RAP93] L. B. Rapoport. On existence of non-smooth invariant functions on the boundary of the absolute stability domain. *Automation and Remote Control*, 54(3):448–453, 1993.

[RB92] E. Rimon and S. Boyd. Efficient distance computation using best ellipsoid fit. *Proceedings of the 1992 IEEE International Symposium on Intelligent Control*, pages 360–365, August 1992.

[RCDP93] M. A. Rotea, M. Corless, D. Da, and I. R. Petersen. Systems with structured uncertainty: Relations between quadratic and robust stability. *IEEE Trans. Aut. Control*, AC-38(5):799–803, 1993.

[REI46] W. T. Reid. A matrix differential equation of the Riccati type. *Amer. J. Math.*, 68:237–246, 1946.

[RIN91] U. T. Ringertz. Optimal design of nonlinear shell structures. Technical Report FFA TN 1991-18, The Aeronautical Research Institute of Sweden, 1991.

[RK88] M. A. Rotea and P. P. Khargonekar. Stabilizability of linear time-varying and uncertain systems. *IEEE Trans. Aut. Control*, 33(9), September 1988.

[RK89] M. A. Rotea and P. P. Khargonekar. Stabilization of uncertain systems with norm bounded uncertainty - a control Lyapunov function approach. *SIAM J. on Optimization*, 27(6):1402–1476, November 1989.

[RK91] M. A. Rotea and P. P. Khargonekar. Generalized $\mathbf{H}_2/\mathbf{H}_\infty$ control via convex optimization. In *Proc. IEEE Conf. on Decision and Control*, volume 3, pages 2719–2720, Brighton, England, June 1991. Also in Proc. IMA Workshop on Robust Control, University of Minnesota, 1993.

[RK93] M. A. Rotea and I. Kaminer. Generalized $\mathbf{H}_2/\mathbf{H}_\infty$ control of discrete-time systems. In *12th IFAC World Congress*, volume 2, pages 239–242, Sydney, Australia, July 1993.

[ROB89] G. Robel. On computing the infinity norm. *IEEE Trans. Aut. Control*, AC-34(8):882–884, August 1989.

[ROC70] R. T. Rockafellar. *Convex Analysis*. Princeton Univ. Press, Princeton, second edition, 1970.

[ROC82] R. T. Rockafellar. *Convex Analysis and Optimization*. Pitman, Boston, 1982.

[Roc93] R. T. Rockafellar. Lagrange multipliers and optimality. *SIAM Review*, 35:183–283, 1993.

[Rot93] M. A. Rotea. The generalized $\mathbf{H}_2$ control problem. *Automatica*, 29(2):373–383, 1993.

[Rox65] E. Roxin. On generalized dynamical systems defined by contingent equations. *Journal of Differential Equations*, 1:188–205, 1965.

[RP93] M. A. Rotea and R. Prasanth. The ρ performance measure: A new tool for controller design with multiple frequency domain specifications. *Submitted to Trans. Aut. Control*, 1993.

[RP94] M. A. Rotea and R. Prasanth. The ρ performance measure: A new tool for controller design with multiple frequency domain specifications. In *Proc. American Control Conf.*, Baltimore, 1994.

[RSH90] S. Richter and A. Scotteward Hodel. Homotopy methods for the solution of general modified algebraic Riccati equations. In *Proc. IEEE Conf. on Decision and Control*, pages 971–976, 1990.

[RVW93] F. Rendl, R. Vanderbei, and H. Wolkowocz. A primal-dual interior-point method for the max-min eigenvalue problem. Technical report, University of Waterloo, Dept. of Combinatorics and Optimization, Waterloo, Ontario, Canada, 1993.

[RW94a] M. A. Rotea and D. Williamson. Optimal realizations of finite wordlength digital controllers via affine matrix inequalities. In *Proc. American Control Conf.*, Baltimore, 1994.

[RW94b] M. A. Rotea and D. Williamson. Optimal synthesis of finite wordlength filters via affine matrix inequalities. *IEEE Trans. Circuits Syst.*, 1994. To appear. Abridged version in Proc. 31st Annual Allerton Conference on Communication, Control, and Computing, Monticello, IL, pp. 505-514, 1993.

[Rya88] E. P. Ryan. Adaptive stabilization of a class of uncertain nonlinear systems: A differential inclusion approach. *Systems and Control Letters*, 10:95–101, 1988.

[SA88] M. Saeki and M. Araki. Search method for the weighting matrix of the multivariable circle criterion. *Int. J. Control*, 47(6):1811–1821, 1988.

[Sae86] M. Saeki. A method of robust stability analysis with highly structured uncertainties. *IEEE Trans. Aut. Control*, AC-31(10):935–940, October 1986.

[Saf80] M. G. Safonov. *Stability and Robustness of Multivariable Feedback Systems*. MIT Press, Cambridge, 1980.

[Saf82] M. G. Safonov. Stability margins of diagonally perturbed multivariable feedback systems. *IEE Proc.*, 129-D:251–256, 1982.

[Saf86] M. G. Safonov. Optimal diagonal scaling for infinity-norm optimization. *Syst. Control Letters*, 7:257–260, 1986.

[Sam59] J. C. Samuels. On the mean square stability of random linear systems. *Trans. IRE*, PGIT-5(Special supplement):248–, 1959.

[San64] I. W. Sandberg. A frequency-domain condition for the stability of feedback systems containing a single time-varying nonlinear element. *Bell Syst. Tech. J.*, 43(3):1601–1608, July 1964.

[San65a] I. W. Sandberg. On the boundedness of solutions of non-linear integral equations. *Bell Syst. Tech. J.*, 44:439–453, 1965.

[San65b] I. W. Sandberg. Some results in the theory of physical systems governed by nonlinear functional equations. *Bell Syst. Tech. J.*, 44:871–898, 1965.

[Sav91] A. V. Savkin. Absolute stability of nonlinear control systems with nonstationary linear part. *Automation and Remote Control*, 41(3):362–367, 1991.

[SC85] A. Steinberg and M. Corless. Output feedback stabilization of uncertain dynamical systems. *IEEE Trans. Aut. Control*, AC-30(10):1025–1027, October 1985.

[SC91] A. G. Soldatos and M. Corless. Stabilizing uncertain systems with bounded control. *Dynamics and Control*, 1:227–238, 1991.

[SC93] M. G. Safonov and R. Y. Chiang. Real/complex K_m-synthesis without curve fitting. In C. T. Leondes, editor, *Control and Dynamic Systems*, volume 56, pages 303–324. Academic Press, New York, 1993.

[SCH68] F. C. Schweppe. Recursive state estimation: Unknown but bounded errors and system inputs. *IEEE Trans. Aut. Control*, AC-13(1):22–28, 1968.

[SCH73] F. C. Schweppe. *Uncertain dynamic systems*. Prentice-Hall, Englewood Cliffs, N.J., 1973.

[SD83] M. G. Safonov and J. Doyle. Optimal scaling for multivariable stability margin singular value computation. In *Proceedings of MECO/EES 1983 Symposium*, 1983.

[SD84] M. G. Safonov and J. Doyle. Minimizing conservativeness of robust singular values. In S. G. Tzafestas, editor, *Multivariable Control*, pages 197–207. D. Reidel, 1984.

[SD86] A. Sideris and R. deGaston. Multivariable stability margin calculation with uncertain correlated parameters. In *Proceedings of 25th Conference on Decision and Control*, pages 766–771. IEEE, December 1986.

[SGL94] M. G. Safonov, K. C. Goh, and J. H. Ly. Control system synthesis via bilinear matrix inequalities. In *Proc. American Control Conf.*, 1994.

[SH68] A. W. Starr and Y. C. Ho. Nonzero-sum differential games. Technical Report 564, Harvard University - Division of Engineering and Applied Physics, Cambridge, Massachusetts, 1968.

[SHA85] A. Shapiro. Extremal problems on the set of nonnegative definite matrices. *Linear Algebra and Appl.*, 67:7–18, 1985.

[SHO85] N. Z. Shor. *Minimization Methods for Non-differentiable Functions*. Springer Series in Computational Mathematics. Springer-Verlag, Berlin, 1985.

[SI93] R. E. Skelton and T. Iwasaki. Liapunov and covariance controllers. *Int. J. Control*, 57(3):519–536, March 1993.

[SIG93] R. E. Skelton, T. Iwasaki, and K. M. Grigoriadis. A unified approach to linear control design, 1993. Manuscript.

[SIL69] D. D. Siljak. *Nonlinear Systems: Parameter Analysis and Design*. John Wiley & Sons, New York, 1969.

[SIL71] D. D. Siljak. New algebraic criteria for positive realness. *J. Franklin Inst.*, 291(2):109–118, February 1971.

[SIL73] D. D. Siljak. Algebraic criteria for positive realness relative to the unit circle. *J. Franklin Inst.*, 296(2):115–122, August 1973.

[SIL89] D. D. Siljak. Parameter space methods for robust control design: A guided tour. *IEEE Trans. Aut. Control*, 34(7):674–688, July 1989.

[SKO90] V. I. Skorodinskii. Iterational method of construction of Lyapunov-Krasovskii functionals for linear systems with delay. *Automation and Remote Control*, 51(9):1205–1212, 1990.

[SKO91A] V. I. Skorodinskii. Iterational methods of testing frequency domain criteria of absolute stability for continuous control systems. I. *Automation and Remote Control*, 52(7):941–945, 1991.

[SKO91B] V. I. Skorodinskii. Iterational methods of verification for frequency domain criteria of absolute stability for continuous control systems. II. *Automation and Remote Control*, 52(8):1082–1088, 1991.

[SKS93] W. Sun, P. Khargonekar, and D. Shim. Controller synthesis to render a closed loop transfer function positive real. In *Proc. American Control Conf.*, volume 2, pages 1074–1078, San Francisco, California, June 1993.

[SL93A] M. G. Safonov and P. H. Lee. A multiplier method for computing real multivariable stability margin. In *IFAC World Congress*, Sydney, Australia, July 1993.

[SL93B] T. J. Su and P. L. Liu. Robust stability for linear uncertain time-delay systems with delay-dependence. *Int. J. Systems Sci.*, 24(6):1067–1080, 1993.

[SLC94] M. G. Safonov, J. Ly, and R. Y. Chiang. On computation of multivariable stability margin using generalized Popov multiplier—LMI approach. In *Proc. American Control Conf.*, Baltimore, June 1994.

[SON83] E. Sontag. A Lyapunov-like characterization of asymptotic controllability. *SIAM J. Contr. Opt.*, 21:462–471, May 1983.

[SON88] G. Sonnevend. New algorithms in convex programming based on a notion of 'centre' (for systems of analytic inequalities) and on rational extrapolation. *International Series of Numerical Mathematics*, 84:311–326, 1988.

[SON90] E. D. Sontag. *Mathematical Control Theory*, volume 6 of *Texts in Applied Mathematics*. Springer-Verlag, 1990.

[SP89] A. Sideris and R. S. S. Peña. Fast computation of the multivariable stability margin for real interrelated uncertain parameters. *IEEE Trans. Aut. Control*, AC-34(12):1272–1276, December 1989.

[SP94A] M. G. Safonov and G. P. Papavassilopoulos. The diameter of an intersection of ellipsoids and BMI robust synthesis. In *Symposium on Robust Control Design*. IFAC, 1994.

[SP94B] A. V. Savkin and I. Petersen. A connection between $\mathbf{H}^\infty$ control and the absolute stabilizability of uncertain systems. *Syst. Control Letters*, 1994. To appear.

[SP94C] A. V. Savkin and I. Petersen. A method for robust stabilization related to the Popov stability criterion. Technical report, Dept. of Elec. Engineering, Australian Defence Force Academy, Campbell 2600, Australia, 1994.

[SP94D] A. V. Savkin and I. Petersen. Minimax optimal control of uncertain systems with structured uncertainty. *Int. J. Robust and Nonlinear Control*, 1994. To appear.

[SP94E] A. V. Savkin and I. Petersen. Nonlinear versus linear control in the absolute stabilizability of uncertain systems with structured uncertainty. *IEEE Trans. Aut. Control*, 1994. Submitted.

[SP94F] A. V. Savkin and I. Petersen. Output feedback ultimate boundedness control for uncertain dynamic systems with structured uncertainty. Technical report, Dept. of Elec. Engineering, Australian Defence Force Academy, Campbell 2600, Australia, 1994.

[SP94G] A. V. Savkin and I. Petersen. Time-varying problems of $\mathbf{H}^\infty$ control and absolute stabilizability of uncertain systems. Technical report, Dept. of Elec. Engineering, Australian Defence Force Academy, Campbell 2600, Australia, 1994. Submitted to 1994 ACC.

[SP94H] A. V. Savkin and I. Petersen. An uncertainty averaging approach to optimal guaranteed cost control of uncertain systems with structured uncertainty. Technical report, Dept. of Elec. Engineering, Australian Defence Force Academy, Campbell 2600, Australia, 1994.

[SRC93] S. M. Swei, M. A. Rotea, and M. Corless. System order reduction in robust stabilization problems. *Int. J. Control*, 1993. To appear; also in 12th IFAC World Congress, Australia 1993, (6)27-30, July, 1993.

[SS72] F. M. Schlaepfer and F. C. Schweppe. Continuous-time state estimation under disturbances bounded by convex sets. *IEEE Trans. Aut. Control*, AC-17(2):197–205, 1972.

[SS81] T. N. Sharma and V. Singh. On the absolute stability of multivariable discrete-time nonlinear systems. *IEEE Trans. Aut. Control*, AC-26(2):585–586, 1981.

[SS90] G. C. Sabin and D. Summers. Optimal technique for estimating the reachable set of a controlled n-dimensional linear system. *Int. J. Systems Sci.*, 21(4):675–692, 1990.

[SSHJ65] D. G. Schultz, F. T. Smith, H. C. Hsieh, and C. D. Johnson. The generation of Liapunov functions. In C. T. Leondes, editor, *Advances in Control Systems*, volume 2, pages 1–64. Academic Press, New York, New York, 1965.

[SSL93] E. Schömig, M. Sznaier, and U. L. Ly. Minimum control effort state-feedback $\mathbf{H}_\infty$-control. In *Proc. American Control Conf.*, volume 1, pages 595–599, San Francisco, California, June 1993.

[ST90] A. Stoorvogel and H. Trentelman. The quadratic matrix inequality in singular $\mathbf{H}_\infty$ control with state feedback. *SIAM J. Control and Optimization*, 28(5):1190–1208, September 1990.

[STO91] A. A. Stoorvogel. The robust $\mathbf{H}_2$ problem: A worst case design. In *Conference on Decision and Control*, pages 194–199, Brighton, England, December 1991.

[STR66] R. L. Stratonovitch. A new representation of stochastic integrals and equations. *SIAM J. Control and Opt.*, pages 362–371, 1966.

[SW87] M. G. Safonov and G. Wyetzner. Computer-aided stability criterion renders Popov criterion obsolete. *IEEE Trans. Aut. Control*, AC-32(12):1128–1131, December 1987.

[SZ92] R. E. Skelton and G. Zhu. Optimal $\mathbf{L}_\infty$ bounds for disturbance robustness. *IEEE Trans. Aut. Control*, AC-37(10):1568–1572, October 1992.

[SZE63] G. P. Szegö. On the absolute stability of sampled-data control systems. *Proc. Nat. Acad. Sci.*, 50:558–560, 1963.

[TB81] J. S. Thorp and B. R. Barmish. On guaranteed stability of uncertain linear systems via linear control. *Journal of Optimization Theory and Applications*, 35(4):559–579, December 1981.

[THI84] L. Thiele. State space discrete systems. *Int. J. of Circuit Theory and Applications*, 12, 1984.

[TM89] H. T. Toivonen and P. M. Mäkilä. Computer-aided design procedure for multiobjective LQG control problems. *Int. J. Control*, 49(2):655–666, February 1989.

[TOI84] H. T. Toivonen. A multiobjective linear quadratic Gaussian control problem. *IEEE Trans. Aut. Control*, AC-29(3):279–280, March 1984.

[TSC92] J. A. Tekawy, M. G. Safonov, and R. Y. Chiang. Convexity property of the one-sided multivariable stability margin. *IEEE Trans. Aut. Control*, AC-37(4):496–498, April 1992.

[TSL92] J. A. Tekawy, M. G. Safonov, and C. T. Leondes. Robustness measures for one-sided parameter uncertainties. *Syst. Control Letters*, 19(2):131–137, August 1992.

[TSY64A] Ya. Tsypkin. Absolute stability of a class of nonlinear automatic sampled data systems. *Aut. Remote Control*, 25:918–923, 1964.

[TSY64B] Ya. Tsypkin. A criterion for absolute stability of automatic pulse systems with monotonic characteristics of the nonlinear element. *Sov. Phys. Doklady*, 9:263–266, 1964.

[TSY64C] Ya. Tsypkin. Frequency criteria for the absolute stability of nonlinear sampled data systems. *Aut. Remote Control*, 25:261–267, 1964.

[TSY64D] Ya. Tsypkin. Fundamentals of the theory of non-linear pulse control systems. In *Proc. 2nd IFAC*, pages 172–180, London, 1964. Butterworth & Co.

[TV91] A. Tesi and A. Vicino. Robust absolute stability of Lur'e control systems in parameter space. *Automatica*, 27(1):147–151, January 1991.

[UHL79] F. Uhlig. A recurring theorem about pairs of quadratic forms and extensions: A survey. *Linear Algebra and Appl.*, 25:219–237, 1979.

[VAV91] S. A. Vavasis. *Nonlinear Optimization: Complexity Issues.* Oxford University Press, New York, 1991.

[VB93A] L. Vandenberghe and S. Boyd. A polynomial-time algorithm for determining quadratic Lyapunov functions for nonlinear systems. In *Eur. Conf. Circuit Th. and Design*, pages 1065–1068, 1993.

[VB93B] L. Vandenberghe and S. Boyd. Primal-dual potential reduction method for problems involving matrix inequalities. To be published in *Math. Programming*, 1993.

[VID78] M. Vidyasagar. On matrix measures and convex Liapunov functions. *J. Math. Anal. and Appl.*, 62:90–103, 1978.

[VID92] M. Vidyasagar. *Nonlinear Systems Analysis.* Prentice-Hall, Englewood Cliffs, second edition, 1992.

[VV85] A. Vannelli and M. Vidyasagar. Maximal Lyapunov functions and domains of attraction for autonomous nonlinear systems. *Automatica*, 21(1):69–80, 1985.

[WAT91] G. A. Watson. An algorithm for optimal ℓ_2 scaling of matrices. *IMA Journal of Numerical Analysis*, 11:481–492, 1991.

[WB71] J. C. Willems and G. L. Blankenship. Frequency domain stability criteria for stochastic systems. *IEEE Trans. Aut. Control*, AC-16(4):292–299, 1971.

[WBL92] M. Wang, D. Boley, and E. B. Lee. Robust stability and stabilization of time delay systems in real parameter space. In *Proc. American Control Conf.*, volume 1, pages 85–86, Chicago, June 1992.

[WEI90] K. Wei. Quadratic stabilizability of linear systems with structural independent time-varying uncertainties. *IEEE Trans. Aut. Control*, AC-35(3):268–277, March 1990.

[WEI94] A. Weinmann. *Uncertain models and robust control.* Springer-Verlag, 1994.

[WEN88] J. T. Wen. Time domain and frequency domain conditions for strict positive realness. *IEEE Trans. Aut. Control*, 33:988–992, 1988.

[WIL69A] J. C. Willems. *The Analysis of Feedback Systems*, volume 62 of *Research Monographs.* MIT Press, 1969.

[WIL69B] J. C. Willems. Stability, instability, invertibility, and causality. *SIAM J. Control*, 7(4):645–671, November 1969.

[WIL70] J. L. Willems. *Stability Theory of Dynamical Systems.* John Wiley & Sons, New York, 1970.

[WIL71A] J. C. Willems. The generation of Lyapunov functions for input-output stable systems. *SIAM J. Control*, 9(1):105–134, February 1971.

[WIL71B] J. C. Willems. Least squares stationary optimal control and the algebraic Riccati equation. *IEEE Trans. Aut. Control*, AC-16(6):621–634, December 1971.

[WIL72] J. C. Willems. Dissipative dynamical systems I: General theory. II: Linear systems with quadratic supply rates. *Archive for Rational Mechanics and Analysis*, 45:321–343, 1972.

[WIL73] J. L. Willems. The circle criterion and quadratic Lyapunov functions for stability analysis. *IEEE Trans. Aut. Control*, 18:184–186, 1973.

[WIL74A] J. C. Willems. On the existence of a nonpositive solution to the Riccati equation. *IEEE Trans. Aut. Control*, AC-19(5), October 1974.

[WIL74B] A. N. Willson Jr. A stability criterion for nonautonomous difference equations with application to the design of a digital FSK oscillator. *IEEE Trans. Circuits Syst.*, 21:124–130, 1974.

[WIL76] J. C. Willems. Mechanisms for the stability and instability in feedback systems. *Proc. IEEE*, 64:24–35, 1976.

[WIL79] A. S. Willsky. *Digital Signal Processing and Control and Estimation Theory: Points of Tangency, Areas of Intersection, and Parallel Directions.* MIT Press, 1979.

[WIL70] A. L. Willson. New theorems on the equations of nonlinear DC transistor networks. *Bell System technical Journal*, 49:1713–1738, October 70.

[WOE90] H. J. Woerdeman. Strictly contractive and positive completions for block matrices. *Linear Algebra and Appl.*, 105:63–105, 1990.

[WON66] W. M. Wonham. Lyapunov criteria for weak stochastic stability. *J. of Diff. Equations*, 2:195–207, 1966.

[WON68] W. M. Wonham. On a matrix Riccati equation of stochastic control. *SIAM J. Control and Opt.*, 6:681–697, 1968.

[WRI92] M. Wright. Interior methods for constrained optimization. *Acta Numerica*, 1, 1992.

[WW83] J. L. Willems and J. C. Willems. Robust stabilization of uncertain systems. *SIAM J. Control and Opt.*, 21(3):352–373, May 1983.

[WXDS92] Y. Wang, L. Xie, and C. E. de Souza. Robust control of a class of uncertain non-linear systems. *Syst. Control Letters*, 19:139–149, 1992.

[XDS92] L. Xie and C. E. de Souza. Robust $\mathbf{H}_\infty$ control for linear systems with norm-bounded time-varying uncertainty. *IEEE Trans. Aut. Control*, AC-37(8), August 1992.

[XFDS92] L. Xie, M. Fu, and C. E. de Souza. $\mathbf{H}_\infty$ control and quadratic stabilization of systems with parameter uncertainty via output feedback. *IEEE Trans. Aut. Control*, 37(8):1253–1256, August 1992.

[YAK62] V. A. Yakubovich. The solution of certain matrix inequalities in automatic control theory. *Soviet Math. Dokl.*, 3:620–623, 1962. In Russian, 1961.

[YAK64] V. A. Yakubovich. Solution of certain matrix inequalities encountered in non-linear control theory. *Soviet Math. Dokl.*, 5:652–656, 1964.

[YAK66] V. A. Yakubovich. Frequency conditions for the existence of absolutely stable periodic and almost periodic limiting regimes of control systems with many nonstationary elements. In *IFAC World Congress*, London, 1966.

[YAK67] V. A. Yakubovich. The method of matrix inequalities in the stability theory of nonlinear control systems, *I*, *II*, *III*. *Automation and Remote Control*, 25-26(4):905–917, 577–592, 753–763, April 1967.

[YAK70] V. A. Yakubovich. On synthesis of optimal controls in a linear differential game with quadratic payoff. *Soviet Math. Dokl.*, 11(6):1478–1481, 1970.

[YAK71] V. A. Yakubovich. On synthesis of optimal controls in a linear differential game on a finite time interval with quadratic payoff. *Soviet Math. Dokl.*, 12(5):1454–1458, 1971.

[YAK73] V. A. Yakubovich. Minimization of quadratic functionals under the quadratic constraints and the necessity of a frequency condition in the quadratic criterion for absolute stability of nonlinear control systems. *Soviet Math. Doklady*, 14:593–597, 1973.

[YAK77] V. A. Yakubovich. The $\mathcal{S}$-procedure in non-linear control theory. *Vestnik Leningrad Univ. Math.*, 4:73–93, 1977. In Russian, 1971.

[YAK82] V. A. Yakubovich. Frequency methods of qualitative investigation into nonlinear systems. In A. Yu. Ishlinsky and F. L. Chernousko, editors, *Advances in Theoretical and Applied Mechanics*, pages 102–124. MIR, Moscow, 1982.

[YAK88] V. A. Yakubovich. Dichotomy and absolute stability of nonlinear systems with periodically nonstationary linear part. *Syst. Control Letters*, 1988.

[YAK92] V. A. Yakubovich. Nonconvex optimization problem: The infinite-horizon linear-quadratic control problem with quadratic constraints. *Syst. Control Letters*, July 1992.

[YAN92] Q. Yang. *Minimum Decay Rate of a Family of Dynamical Systems*. PhD thesis, Stanford University, Stanford, CA 94305, January 1992.

[YAZ93] E. Yaz. Stabilizing compensator design for uncertain nonlinear systems. *Syst. Control Letters*, 21(1):11–17, 1993.

[YOS94] A. Yoshise. An optimization method for convex programs—interior-point method and analytical center. *Systems, Control and Information*, 38(3):155–160, March 1994. Special issue on Numerical Approaches in Control Theory. In Japanese.

[ZAM66A] G. Zames. On the input-output stability of time-varying nonlinear feedback systems—Part I: Conditions derived using concepts of loop gain, conicity, and positivity. *IEEE Trans. Aut. Control*, AC-11:228–238, April 1966.

[ZAM66B] G. Zames. On the input-output stability of time-varying nonlinear feedback systems—Part II: Conditions involving circles in the frequency plane and sector nonlinearities. *IEEE Trans. Aut. Control*, AC-11:465–476, July 1966.

[ZD63] L. A. Zadeh and C. A. Desoer. *Linear System Theory: The State Space Approach.* McGraw-Hill, New York, 1963. Reprinted by Krieger, 1979.

[ZF67] G. Zames and P. L. Falb. On the stability of systems with monotone and odd monotone nonlinearities. *IEEE Trans. Aut. Control*, 12(2):221–223, April 1967.

[ZF68] G. Zames and P. L. Falb. Stability conditions for systems with monotone and slope-restricted nonlinearities. *SIAM J. Control*, 6(1):89–108, July 1968.

[ZF83] G. Zames and B. A. Francis. Feedback, minimax sensitivity and optimal robustness. *IEEE Trans. Aut. Control*, AC-28(5):585–601, May 1983.

[ZIN90] A. S. I. Zinober, editor. *Deterministic control of uncertain systems.* Peter Peregrinus Ltd., London, 1990.

[ZK87] K. Zhou and P. P. Khargonekar. Stability robustness bounds for linear state-space models with structured uncertainty. *IEEE Trans. Aut. Control*, AC-32(7):621–623, July 1987.

[ZK88] K. Zhou and P. P. Khargonekar. Robust stabilization of linear systems with norm-bounded time-varying uncertainty. *Syst. Control Letters*, 10:17–20, 1988.

[ZKSN92] K. Zhou, P. P. Khargonekar, J. Stoustrup, and H. H. Niemann. Robust stability and performance of uncertain systems in state space. In *Proc. IEEE Conf. on Decision and Control*, pages 662–667, Tucson, December 1992.

[ZRS93] G. Zhu, M. A. Rotea, and R. E. Skelton. A convergent feasible algorithm for the output covariance constraint problem. In *Proc. American Control Conf.*, pages 1675–1679, San Fransico, June 1993.

[ZUB55] V. I. Zubov. Problems in the theory of the second method of Lyapunov, construction of the general solution in the domain of asymptotic stability. *Prikladnaya Matematika i Mekhanika*, 19:179–210, 1955.

Index